T0340179

IMAGING GENETICS

THE ELSEVIER AND MICCAI SOCIETY BOOK SERIES

Also available:
Wu, Machine Learning and Medical Imaging,
9780128040768
Zhou, Medical Image Recognition, Segmentation and Parsing,
9780128025819
Zhou, Deep Learning for Medical Image Analysis,
9780128104088

IMAGING GENETICS

Edited by

ADRIAN V. DALCA

CSAIL, Mass. Institute of Technology; and Postdoctoral Fellow Martinos Center for Biomedical Imaging, Mass. General Hospital, Harvard Medical School

NEMATOLLAH K. BATMANGHELICH

Assistant Professor, Department of Biomedical Informatics Intelligent Systems Program, University of Pittsburgh Pittsburgh

LI SHEN

Associate Professor of Radiology and Imaging Sciences Center for Neuroimaging, Radiology and Imaging Sciences, Indiana University School of Medicine, Indianapolis, Indiana

MERT R. SABUNCU

Assistant Professor, Electrical and Computer Engineering Biomedical Engineering, Cornell University

Academic Press is an imprint of Elsevier
125 London Wall, London EC2Y 5AS, United Kingdom
525 B Street, Suite 1800, San Diego, CA 92101-4495, United States
50 Hampshire Street, 5th Floor, Cambridge, MA 02139, United States
The Boulevard, Langford Lane, Kidlington, Oxford OX5 1GB, United Kingdom

Library of Congress Cataloging-in-Publication Data
A catalog record for this book is available from the Library of Congress

British Library Cataloguing-in-Publication Data
A catalogue record for this book is available from the British Library

ISBN: 978-0-12-813968-4

For information on all Academic Press publications visit our website at https://www.elsevier.com/books-and-journals

Publisher: Mara E. Conner
Acquisition Editor: Tim Pitts
Editorial Project Manager: Anna Valutkevich
Production Project Manager: Mohanapriyan Rajendran
Designer: Matthew Limbert

Typeset by TNQ Books and Journals

CONTENTS

LIST OF CONTRIBUTORS

Hieab H.H. Adams
Erasmus Medical Center, Rotterdam, The Netherlands

Alejandro Arias-Vasquez
Radboud University Medical Center, Nijmegen, The Netherlands

Gareth Ball
King's College London, London, United Kingdom

James P. Boardman
University of Edinburgh, Edinburgh, United Kingdom

Vince D. Calhoun
Mind Research Network, Albuquerque, NM, United States; The University of New Mexico, Albuquerque, NM, United States

Feng Chen
Harbin Engineering University, Harbin, China

Serena J. Counsell
King's College London, London, United Kingdom

Su-Ping Deng
Tulane University, New Orleans, LA, United States; Tongji University, Shanghai, China

Sylvane Desrivieres
King's College London, London, United Kingdom

Greig I. de Zubicaray
Queensland University of Technology (QUT), Brisbane, QLD, Australia

Vincent Doré
Australian eHealth Research Centre, CSIRO, Herston, QLD, Australia

Lei Du
Indiana University School of Medicine, Indianapolis, IN, United States

David Edwards
King's College London, London, United Kingdom

Joshua Faskowitz
Keck School of Medicine of USC, Marina del Rey, CA, United States

Weixing Feng
Harbin Engineering University, Harbin, China

Barbara Franke
Radboud University Medical Center, Nijmegen, The Netherlands

Jurgen Fripp
Australian eHealth Research Centre, CSIRO, Herston, QLD, Australia

Boris A. Gutman
Keck School of Medicine of USC, Marina del Rey, CA, United States

Derrek P. Hibar
Keck School of Medicine of USC, Marina del Rey, CA, United States

De-Shuang Huang
Tongji University, Shanghai, China

Heng Huang
University of Texas at Arlington, Arlington, TX, United States

M. Arfan Ikram
Erasmus Medical Center, Rotterdam, The Netherlands

Alex Ing
King's College London, London, United Kingdom

Mark Inlow
Rose-Hulman Institute of Technology, Terre Haute, IN, United States

Neda Jahanshad
Keck School of Medicine of USC, Marina del Rey, CA, United States

Sungeun Kim
Indiana University School of Medicine, Indianapolis, IN, United States

Rebecca C. Knickmeyer
University of North Carolina at Chapel Hill, Chapel Hill, NC, United States

Michelle L. Krishnan
King's College London, London, United Kingdom

Jin Li
Harbin Engineering University, Harbin, China

Hong Liang
Harbin Engineering University, Harbin, China; Indiana University School of Medicine, Indianapolis, IN, United States

Dongdong Lin
Mind Research Network, Albuquerque, NM, United States

Zhaohua Lu
Pennsylvania State University, State College, PA, United States

Nicholas G. Martin
Queensland Institute of Medical Research, Brisbane, QLD, Australia

Katie L. McMahon
University of Queensland, Brisbane, QLD, Australia

Sarah E. Medland
QIMR Berghofer Medical Research Institute, Brisbane, QLD, Australia

Xianglian Meng
Harbin Engineering University, Harbin, China; Habin Huade University, Harbin, China

Giovanni Montana
King's College London, London, United Kingdom

Jason H. Moore
University of Pennsylvania, Philadelphia, PA, United States

Wiro J. Niessen
Erasmus Medical Center, Rotterdam, The Netherlands

Daniel A. Rinker
Keck School of Medicine of USC, Marina del Rey, CA, United States

Shannon L. Risacher
Indiana University School of Medicine, Indianapolis, IN, United States

Stephen Rose
Australian eHealth Research Centre, CSIRO, Herston, QLD, Australia

Gennady Roshchupkin
Erasmus Medical Center, Rotterdam, The Netherlands

Olivier Salvado
Australian eHealth Research Centre, CSIRO, Herston, QLD, Australia

Andrew J. Saykin
Indiana University School of Medicine, Indianapolis, IN, United States; Habin Huade University, Harbin, China

Gunter Schumann
King's College London, London, United Kingdom

Kaikai Shen
Australian eHealth Research Centre, CSIRO, Herston, QLD, Australia

Li Shen
Indiana University School of Medicine, Indianapolis, IN, United States; Indiana University Indianapolis, Indianapolis, IN, United States

Matt Silver
London School of Hygiene and Tropical Medicine, London, United Kingdom

Paul M. Thompson
Keck School of Medicine of USC, Marina del Rey, CA, United States; University of Southern California, Marina del Rey, CA, United States

Meike W. Vernooij
Erasmus Medical Center, Rotterdam, The Netherlands

Andrew J. Walley
Imperial College London, London, United Kingdom

Zi Wang
Imperial College London, London, United Kingdom

Yu-Ping Wang
Tulane University, New Orleans, LA, United States

Lei Wang
Harbin Engineering University, Harbin, China

Margaret J. Wright
University of Queensland, Brisbane, QLD, Australia; Queensland Institute of Medical Research, Brisbane, QLD, Australia

Jingwen Yan
Indiana University School of Medicine, Indianapolis, IN, United States; Indiana University Indianapolis, Indianapolis, IN, United States; Indiana University School of Informatics and Computing, Indianapolis, IN, United States

Xiaohui Yao
Indiana University School of Medicine, Indianapolis, IN, United States; Indiana University School of Informatics and Computing, Indianapolis, IN, United States

Qiushi Zhang
Harbin Engineering University, Harbin, China; Northeast Dianli University, Jilin, China

Yize Zhao
Weill Cornell Medicine, New York, NY, United States

Hongtu Zhu
University of Texas MD Anderson Cancer Center, Houston TX, United States

Fei Zou
University of Florida, Gainesville, FL, United States

Marcel P. Zwiers
Radboud University Medical Center, Nijmegen, The Netherlands

BIOGRAPHY

Adrian V. Dalca is a postdoctoral fellow at Massachusetts General Hospital, Harvard Medical School, as well as Massachusetts Institute of Technology (MIT). He obtained his PhD from MIT in the Electrical Engineering and Computer Science department. He is interested in mathematical models and machine learning for medical image analysis, with a focus on characterizing genetic and clinical effects on imaging phenotypes. He is also interested and active in healthcare entrepreneurship and translation of algorithms to the clinic.

Mert Sabuncu is an Assistant Professor in Electrical and Computer Engineering, with a secondary appointment in Biomedical Engineering, Cornell University. His research interests are in biomedical data analysis, in particular imaging data, and with an application emphasis on neuroscience and neurology. He uses tools from signal/image processing, probabilistic modeling, statistical inference, computer vision, computational geometry, graph theory, and machine learning to develop algorithms that allow learning from large-scale biomedical data.

Kayhan Batmanghelich is an Assistant Professor of department of Biomedical Informatics and Intelligent Systems Program at the University of Pittsburgh and an adjunct faculty in the Machine Learning department at the Carnegie Mellon University. His research is at the intersection of medical vision, machine learning, and bioinformatics. He develops algorithms to analyze and understand medical image along with genetic data and other electrical health records such as the clinical report. He is interested in method development as well as translational clinical problems.

Li Shen received a BS degree from Xi'an Jiaotong University, an MS degree from Shanghai Jiaotong University, and a PhD degree from Dartmouth College, all in Computer Science. He is an Associate Professor of Radiology and Imaging Sciences at Indiana University School of Medicine. His research interests include medical image computing, bioinformatics, machine learning, network science, brain imaging genomics, and big data science in biomedicine.

LIST OF FIGURES

INTRODUCTION

Adrian V. Dalca, Nematollah K. Batmanghelich, Mert R. Sabuncu, Li Shen

Imaging Genetics studies the relationships between genetic variation and biomedical imaging measurements, often in the context of disease. Facilitated by recent advances in genotyping and imaging technologies, along with large-scale collaborative data collection efforts, imaging genetics is a rapidly growing research field that promises new avenues for biological discovery and personalized medicine but simultaneously faces several unique challenges.

In this chapter, we aim to provide an introduction to this field, concentrating on methodological developments and highlighting some prior studies, while pointing out those included in the following chapters of this book. For a more detailed literature overview, we refer the reader to dedicated survey papers, such as [1–4].

This book is an edited collection of peer-reviewed articles presented at the second MICGen: MICCAI Workshop on Imaging Genetics, held in conjunction with the International Conference on Medical Image Computing and Computer Aided Intervention, MICCAI, 2015.

1. METHODS IN IMAGING GENETICS

Imaging genetics analyzes relationships between genetic or genomic data and measurements extracted from biomedical images. In general, genetic and genomic data can involve direct nucleotide measurements, population level statistics such as minor allele frequencies, gene expression data, and epigenetic markers. Image-derived phenotypes can range from voxel-wise intensities [5] to high-level measurements from specific regions of interest [6], identified manually or automatically. In addition to these modalities, imaging genetics studies can also incorporate other types of data, such as environmental factors or clinical variables [7,8].

Classically, a phenotype is an observable quantitative trait such as a clinical test score or physical measurement or a categorical variable such as disease diagnosis. Imaging genetics employs imaging-based biomarkers as intermediate phenotypes (or endophenotypes) that provide a rich quantitative characterization of disease. This strategy can offer a more complete picture of underlying genetic mechanisms and promises to aid in identifying and confirming relevant genetic variations and pertinent imaging features [1,2,4,9,10].

1.1 Large-Scale Univariate Analyses

Imaging genetics presents substantial methodological challenges due to the extremely high dimensional nature of the data (involving thousands of image voxels and millions of genetic variants) [10–13]. Furthermore, many of the diseases of interest exhibit significant clinical heterogeneity and complex multivariate patterns of association, which, in turn, necessitates strategies such as subphenotyping and refinement of genetic networks [14–16]. In recent years, several methods addressing these challenges have been featured in a wide range of venues, from academic journals, biological conferences to machine learning workshops, underscoring the diversity of the field and the complexity of the problems that are tackled.

Techniques in imaging genetics are often adopted from population genetics, which explores the relationships between genetic variation and general phenotypes. Over the last decade, population genetics has been dominated by genome-wide association studies, where the standard approach has involved interrogating correlations between each genetic marker and each phenotype independently via (univariate) statistical tests. In early imaging genetics studies, univariate tests were also employed by only considering a few candidate imaging variables and/or a handful of genetic markers. The main drawback of this strategy, however, is that it discards most of the data.

Large-scale univariate analyses often require relatively large datasets, such as the Alzheimer's Disease Neuroimaging Initiative, where one main objective is to establish the role of genetic risk factors in Alzheimer's disease by collecting data from several thousand individuals [4,8,17]. Another large-scale study, ENIGMA [18], employs a metaanalytic approach to correlate standard neuroimaging measurements and common genetic markers. CHARGE [19], ISGC, [20] and METASTROKE [21] studies explore genetic effects in stroke, while COPDgene [22] is focused on understanding the etiology of Chronic Obstructive Pulmonary Disease. Recent extensions of the univariate approach include the genome-wide voxel-wise analysis and its variants, where each voxel is treated as an independent phenotype [5,23]. Statistical power is the core limitation in these approaches because conservative significance levels need to be imposed due to the immense number of statistical tests that are conducted. Furthermore, all univariate methods ignore a significant portion of the structure in high dimensional data and thus constrain the discovery potential.

1.2 Multivariate Methods

Multivariate analyses that attempt to capture more complex relationships between variants and phenotypes have recently gained popularity [9,24–27]. For example, a common approach is to a priori identify a small set of image-based features associated with some disease, and subsequently, finds the relevant genetic markers that explain the selected features through a multivariate model [10,28]. Other algorithms involve joint models of various modalities, and the use of approximate Bayesian inference machinery [9,24,26]. A variety of sparse canonical correlation analysis and similar models are also proposed to identify bimultivariate association between imaging and genetic features [25,27,29–31].

Another growing body of work focuses on the optimization and selection of the anatomical features for genetic association analyses [32–34]. For example, one can use established loci associated to a disease to develop better imaging biomarkers, such as using a variant of an admixture model to discover genetically driven subtypes of emphysema [32].

An alternative direction deals with individual-level prediction based on imaging and/or genetic data. In contrast to studies aimed at biological discovery, these approaches use data, such as genetic variables and clinical indicators, to facilitate prognostication. One such example is the future prediction of anatomical changes captured by brain imaging [7,35].

2. IMAGING GENETICS DISCOVERY

Imaging genetics methodologies have led to the identification of novel-causal genetic variants and disease-relevant imaging features. For example, common genetic variants have been demonstrated to be associated with hippocampal volume and intracranial volume in the context of several neurodegenerative diseases [3,4,8,36–38]; obesity and brain volume [39]; and subcortical brain structures [40]. Genetic predispositions are being investigated for small vessel disease in stroke patients [41,42]. The growing availability of data, advances in complex structural and functional models, and improvements in our genetic understanding will continue to fuel novel discovery in imaging genetics studies.

3. CURRENT RESEARCH TRENDS

We discuss some current research trends that are touched on in the chapters of this book. We note that these topics do not represent an exhaustive coverage of the field as imaging genetics has been adopted in a broad range of areas, for example, protein network analysis, blood biomarker discovery [43], heritability analysis [33,44–46], connectome genetics [13], or modeling the environment and its interactions [7,47].

3.1 Joint Modeling of Multiple Modalities

Imaging genetics studies can comprise a wide variety of data, including multimodal imaging, various genetic and epigenetic markers, pathways, clinical variables, and diagnostic status. These heterogeneous sources of information have complex relationships and are often multidimensional. Univariate statistical models and many widely used inference algorithms do not capture these complex interactions. Jointly representing and interrogating these data in a common statistical framework can therefore be very powerful. Chapters 4, 6–9 present sophisticated studies that implement this approach.

3.2 Clinical Application and Translation

While a core goal in imaging genetics is to discover novel genes, some recent studies are tackling complicated clinical questions, which requires new methodologies. For example, imaging and genetics data can be used to subphenotype complex disorders and dissect relevant genetic networks. Chapters 1–3 present examples of the application of these models to large-scale studies.

3.3 Discovery of Relevant Imaging Features

A core objective of biomedical imaging is to identify anatomical correlates of disease, which can be computed via segmentation or other image processing algorithms. One approach in neuroimaging genetics involves developing methods that utilize the genetic signature of disease to identify novel imaging biomarkers. Chapters 1, 2, 5, and 8 present important examples of this type of research.

4. MICGEN: WORKSHOP ON IMAGING GENETICS

The MICGen workshop, which produced the chapters in this book, was organized as part of the International Conference on Medical Image Computing and Computer Aided Intervention (MICCAI) 2015. The workshop aimed to capture the emerging field of imaging genetics and encourage discussion on both fundamental concepts and novel methods. We believe MICCAI offers an ideal venue to bring together researchers with different expertise and shared interests in this rapidly evolving field. Specifically, the mathematical, computer science and engineering experience of the MICCAI community can help develop new methods for the analysis of emerging imaging and genetics data [48–50]. We intend to continue the MICGen tradition and look forward to organizing this workshop at MICCAI 2017.

4.1 MICGen 2015 Organizing Committee

- Adrian V. Dalca, CSAIL, Massachusetts Institute of Technology
- Nematollah K. Batmanghelich, Department of Biomedical Informatics, Intelligent Systems Program, University of Pittsburgh
- Mert R. Sabuncu, Electrical and Computer Engineering, Biomedical Engineering, Cornell University
- Li Shen, Center for Neuroimaging, Radiology and Imaging Sciences, Indiana University School of Medicine

4.2 MICGen 2015 Referees

- Liana Apostolova, Neurology, Radiology and Medical and Molecular Genetics, Indiana University School of Medicine
- Tian Ge, A.A Martinos Center for Biomedical Imaging, Massachusetts General Hospital, Harvard Medical School
- Peter Kochunov, Maryland Psychiatric Research Center
- Albert Montillo, Radiology, Advanced Imaging Research Center
- Bertrand Thirion, Inria, Paris-Saclay University
- Jessica Turner, Psychology, College of Arts & Sciences, Georgia State University
- Yu-Ping Wang, Department of Biomedical Engineering, Tulane University

- Anderson Winkler, Oxford Centre for Functional MRI of the Brain, University of Oxford
- Rachel Yotter, Section of Biomedical Image Analysis, University of Pennsylvania

4.3 MICGen 2014 and 2015 Keynote Speakers

- Liana Apostolova, Neurology, Radiology and Medical and Molecular Genetics, Indiana University School of Medicine
- Mark Daly, Massachusetts General Hospital, Harvard Medical School
- Derrek Hibar, Laboratory of Neuro Imaging, University of Southern California
- Gabi Kastenmuller, Institute of Bioinformatics and Systems Biology, Helmholtz Zentrum Muenchen
- Manolis Kellis, EECS, Massachusetts Institute of Technology
- Burkhard Rost, Computational Biology & Bioinformatics, Department of Informatics, Technical
- Mert Sabuncu, A.A Martinos Center for Biomedical Imaging, Massachusetts General Hospital, Harvard Medical School
- Li Shen, Radiology and Imaging Sciences, Indiana University School of Medicine
- Jordan Smoller, Center for Human Genetic Research, Massachusetts General Hospital University of Munich; President of International Society for Computational Biology, 2007–14

REFERENCES

[1] R. Bogdan, B.J. Salmeron, C.E. Carey, A. Agrawal, V.D. Calhoun, H. Garavan, A.R. Hariri, A. Heinz, M.N. Hill, A. Holmes, et al., Imaging genetics and genomicsin psychiatry: a critical review of progress and potential, Biological Psychiatry 82 (2017).

[2] J. Liu, V.D. Calhoun, A review of multivariate analyses in imaging genetics, Frontiers in Neuroinformatics 8 (2014).

[3] A.J. Saykin, L. Shen, X. Yao, S. Kim, K. Nho, S.L. Risacher, V.K. Ramanan, T.M. Foroud, K.M. Faber, N. Sarwar, L.M. Munsie, X. Hu, H.D. Soares, S.G. Potkin, P.M. Thompson, J.S. Kauwe, R. Kaddurah-Daouk, R.C. Green, A.W. Toga, M.W. Weiner, Initiative Alzheimer's Disease Neuroimaging, Genetic studies of quantitative MCI and AD phenotypes in ADNI: progress, opportunities, and plans, Alzheimer's & Dementia: The Journal of the Alzheimer's Association 11 (7) (2015) 792–814.

[4] L. Shen, P.M. Thompson, S.G. Potkin, L. Bertram, L.A. Farrer, T.M. Foroud, R.C. Green, X. Hu, M.J. Huentelman, S. Kim, et al., Genetic analysis of quantitative phenotypes in AD and MCI: imaging, cognition and biomarkers, Brain Imaging and Behavior 8 (2) (2014) 183–207.

[5] J.L. Stein, X. Hua, S. Lee, A.J. Ho, A.D. Leow, A.W. Toga, A.J. Saykin, L. Shen, T. Foroud, N. Pankratz, et al., Voxelwise genome-wide association study (VGWAS), Neuroimage 53 (3) (2010) 1160–1174.
[6] L. Shen, S. Kim, S.L. Risacher, K. Nho, S. Swaminathan, J.D. West, T. Foroud, N. Pankratz, J.H. Moore, C.D. Sloan, M.J. Huentelman, D.W. Craig, B.M. Dechairo, S.G. Potkin Jr., C.R. Jack, M.W. Weiner, A.J. Saykin, Initiative Alzheimer's Disease Neuroimaging, Whole genome association study of brain-wide imaging phenotypes for identifying quantitative trait loci in mci and ad: a study of the adnicohort, Neuroimage 53 (3) (2010) 1051–1063.
[7] A.V. Dalca, R. Sridharan, M.R. Sabuncu, P. Golland, Predictive modeling of anatomy with genetic and clinical data, MICCAI: International Conference on Medical Image Computing and Computer Assisted Intervention, LNCS 9351 (2015) 519–526.
[8] M.W. Weiner, D.P. Veitch, P.S. Aisen, L.A. Beckett, N.J. Cairns, R.C. Green, D. Harvey, C.R. Jack, W. Jagust, E. Liu, et al., The alzheimer's disease neuroimaging initiative: a review of papers published since its inception, Alzheimer's & Dementia 9 (5) (2013) e111–e194.
[9] N. Batmanghelich, A.V. Dalca, G. Quon, M. Sabuncu, P. Golland, Probabilistic modeling of imaging, genetics and diagnosis, IEEE Transactions on Medical Imaging (2016).
[10] M. Vounou, T.E. Nichols, G. Montana, Discovering genetic associations with high-dimensional neuroimaging phenotypes:a sparse reduced-rank regression approach, Neuroimage 53 (3) (2010) 1147–1159.
[11] S. Mackey, K.-J. Kan, B. Chaarani, N. Alia-Klein, A. Batalla, S. Brooks, J. Cousijn, A. Dagher, M. Ruiter, S. Desrivieres, et al., Genetic imaging consortium for addiction medicine: from neuroimaging to genes, Progress in Brain Research 224 (2016) 203–223.
[12] J.-B. Poline, C. Lalanne, A. Tenenhaus, E. Duchesnay, B. Thirion, V. Frouin, Imaging genetics: bio-informatics and bio-statistics challenges, Proceedings of COMP-stat'2010 (2010) 101–110.
[13] P.M. Thompson, T. Ge, D.C. Glahn, N. Jahanshad, T.E. Nichols, Genetics of the connectome, Neuroimage 80 (2013) 475–488.
[14] A.P. Morris, C.M. Lindgren, E. Zeggini, N.J. Timpson, T.M. Frayling, A.T. Hattersley, M.I. Mc-Carthy, A powerful approach to sub-phenotype analysis in population-based genetic association studies, Genetic Epidemiology 34 (4) (2010) 335–343.
[15] A.L. Price, N.J. Patterson, R.M. Plenge, M.E. Weinblatt, N.A. Shadick, D. Reich, Principal components analysis corrects for stratification in genome-wide association studies, Nature Genetics 38 (8) (2006) 904–909.
[16] K. Wang, M. Li, H. Hakonarson, Analysing biological pathways in genome-wide association studies, Nature Reviews Genetics 11 (12) (2010) 843–854.
[17] C.R. Jack, M.A. Bernstein, N.C. Fox, P. Thompson, G. Alexander, D. Harvey, B. Borowski, P.J. Britson, J.L. Whitwell, C. Ward, et al., The Alzheimer's disease neuroimaging initiative (ADNI): MRI methods, Journal of Magnetic Resonance Imaging 27 (4) (2008) 685–691.
[18] P.M. Thompson, J.L. Stein, S.E. Medland, D.P. Hi-bar, A.A. Vasquez, M.E. Renteria, R. Toro, N. Jahanshad, G. Schumann, B. Franke, et al., The enigma consortium: large-scale collaborative analyses of neuroimaging and genetic data, Brain Imaging and Behavior 8 (2) (2014) 153–182.
[19] B.M. Psaty, C.J. O'Donnell, V. Gudnason, K.L. Lunetta, A.R. Folsom, J.I. Rotter, A.G. Uitter-linden, T.B. Harris, J.C.M. Witteman, E. Boerwinkle, et al., Cohorts for heart and aging research in genomic epidemiology (charge) consortium design of prospective meta-analyses of genome-wide association studies from 5 cohorts, Circulation: Cardiovascular Genetics 2 (1) (2009) 73–80.

[20] J.F. Meschia, D.K. Arnett, H. Ay, R.D. Brown, O.R. Benavente, J.W. Cole, P.I.W. De Bakker, M. Dichgans, K.F. Doheny, M. Fornage, et al., Stroke genetics network (sign) study design and rationale for a genome-wide association study of ischemic stroke subtypes, Stroke 44 (10) (2013) 2694–2702.
[21] M. Traylor, M. Farrall, E.G. Holliday, C. Sudlow, J.C. Hopewell, Y.-C. Cheng, M. Fornage, M.A. Ikram, R. Malik, S. Bevan, et al., Genetic risk factors for ischaemic stroke and its subtypes (the metastroke collaboration): a meta-analysis of genome-wide association studies, The Lancet Neurology 11 (11) (2012) 951–962.
[22] E.A. Regan, J.E. Hokanson, J.R. Murphy, B. Make, D.A. Lynch, T.H. Beaty, D. Curran-Everett, E.K. Silverman, J.D. Crapo, Genetic epidemiology of COPD (COPDgene) study design, COPD: Journal of Chronic Obstructive Pulmonary Disease 7 (1) (2011) 32–43.
[23] D.P. Hibar, J.L. Stein, O. Kohannim, N. Jahanshad, A.J. Saykin, L. Shen, S. Kim, N. Pankratz, T. Foroud, M.J. Huentelman, et al., Voxelwise gene-wide association study (vGeneWAS): multivariate gene-based association testing in 731 elderly subjects, Neuroimage 56 (4) (2011) 1875–1891.
[24] N.K. Batmanghelich, A.V. Dalca, M.R. Sabuncu, P. Golland, Joint modeling of imaging and genetics, Information Processing in medical imaging, LNCS 7917 (2013) 766–777.
[25] L. Du, Y. Jingwen, S. Kim, S.L. Risacher, H. Huang, M. Inlow, J.H. Moore, A.J. Saykin, L. Shen, Initiative Alzheimer's Disease Neuroimaging, A novel structure-aware sparse learning algorithm for brain imaging genetics, Medical Image Computing and Computer-Assisted Intervention: MICCAI 17 (Pt 3) (2014) 329–336.
[26] F.C. Stingo, M. Guindani, M. Vannucci, V.D. Calhoun, An integrative bayesian modeling approach to imaging genetics, Journal of the American Statistical Association 108 (2013).
[27] J. Yan, L. Du, S. Kim, S.L. Risacher, H. Huang, J.H. Moore, A.J. Saykin, L. Shen, Initiative Alzheimer's Disease Neuroimaging, Transcriptome-guided amyloid imaging genetic analysis via a novel structured sparse learning algorithm, Bioinformatics 30 (17) (2014) i564–i571.
[28] M. Vounou, E. Janousova, R. Wolz, J.L. Stein, P.M. Thompson, D. Rueckert, G. Montana, Alzheimer's Disease Neuroimaging Initiative, et al., Sparse reduced-rank regression detects genetic associations with voxel-wise longitudinal phenotypes in alzheimer's disease, Neuroimage 60 (1) (2012) 700–716.
[29] L. Du, H. Huang, J. Yan, S. Kim, S.L. Risacher, M. Inlow, J.H. Moore, A.J. Saykin, L. Shen, Initiative Alzheimer's Disease Neuroimaging, Structured sparse canonical correlation analysis for brain imaging genetics: an improved graphnet method, Bioinformatics 32 (10) (2016) 1544–1551.
[30] X. Hao, C. Li, L. Du, X. Yao, J. Yan, S.L. Risacher, A.J. Saykin, L. Shen, D. Zhang, Initiative Alzheimer's Disease Neuroimaging, Mining outcome-relevant brain imaging genetic associations via three-way sparse canonical correlation analysis in alzheimer's disease, Scientific Reports 7 (2017) 44272.
[31] J. Yan, S.L. Risacher, K. Nho, A.J. Saykin, L.I. Shen, Identification of discriminative imaging proteomics associations in alzheimer's disease via a novel sparse correlation model, Pacific Symposium on Biocomputing 22 (2017) 94–104.
[32] N.K. Batmanghelich, A. Saeedi, M. Cho, R.S.J. Estepar, P. Golland, Generative method to discover genetically driven image biomarkers, in: International Conference on Information Processing and Medical Imaging vol. 9123, 2015, pp. 30–42.
[33] T. Ge, T.E. Nichols, P.H. Lee, A.J. Holmes, J.L. Roffman, R.L. Buckner, M.R. Sabuncu, J.W. Smoller, Massively expedited genome-wide heritability analysis (MEGHA), Proceedings of the National Academy of Sciences 112 (8) (2015) 2479–2484.

[34] A.M. Winkler, P. Kochunov, J. Blangero, L. Almasy, K. Zilles, P.T. Fox, R. Duggirala, D.C. Glahn, Cortical thickness or grey matter volume? the importance of selecting the phenotype for imaging genetics studies, Neuroimage 53 (3) (2010) 1135–1146.
[35] A.V. Dalca, Genetic, Clinical and Population Priors for Brain Images (Ph.D. thesis), Massachusetts Institute of Technology, 2016.
[36] J.L. Stein, S.E. Medland, A.A. Vasquez, D.P. Hibar, R.E. Senstad, A.M. Winkler, R. Toro, K. Ap-pel, R. Bartecek, Ø. Bergmann, et al., Identification of common variants associated with human hippocampal and intracranial volumes, Nature Genetics 44 (5) (2012) 552–561.
[37] Enhancing Neuro Imaging Genetics through Meta, et al., Common variants at 12q14 and 12q24 are associated with hippocampal volume, Nature Genetics 44 (5) (2012) 545–551.
[38] E.T. Westlye, A. Lundervold, H. Rootwelt, A.J. Lundervold, L.T. Westlye, Increased hippocampal default mode synchronization during rest in middle-aged and elderly apoe ε4 carriers: relationships with memory performance, Journal of Neuroscience 31 (21) (2011) 7775–7783.
[39] A.J. Ho, J.L. Stein, X. Hua, S. Lee, D.P. Hibar, A.D. Leow, I.D. Dinov, A.W. Toga, A.J. Saykin, L. Shen, et al., A commonly carried allele of the obesity-related FTO gene is associated with reduced brain volume in the healthy elderly, Proceedings of the National Academy of Sciences 107 (18) (2010) 8404–8409.
[40] D.P. Hibar, J.L. Stein, M.E. Renteria, A. Arias-Vasquez, S. Desrivieres, N. Jahanshad, R. Toro, K. Wittfeld, L. Abramovic, M. Andersson, et al., Common genetic variants influence human subcortical brain structures, Nature 520 (7546) (2015) 224–229.
[41] A. Biffi, C.D. Anderson, J.M. Jagiella, H. Schmidt, B. Kissela, B.M. Hansen, J. Jimenez-Conde, C.R. Pires, A.M. Ayres, K. Schwab, et al., A poe genotype andextent of bleeding and outcome in lobar intracerebral haemorrhage: agenetic association study, The Lancet Neurology 10 (8) (2011) 702–709.
[42] N.S. Rost, S.M. Greenberg, J. Rosand, The genetic architecture of intracerebral hemorrhage, Stroke 39 (7) (2008) 2166–2173.
[43] S. Kim, S. Swaminathan, L. Shen, S.L. Risacher, K. Nho, T. Foroud, L.M. Shaw, J.Q. Trojanowski, S.G. Potkin, M.J. Huentelman, et al., Genome-wide association study of CSF biomarkers aβ1-42, t-tau, and p-tau181pin the ADNI cohort, Neurology 76 (1) (2011) 69–79.
[44] S. Durston, P. de Zeeuw, W.G. Staal, Imaging genetics in ADHD: a focus on cognitive control, Neuroscience & Biobehavioral Reviews 33 (5) (2009) 674–689.
[45] T. Ge, A.J. Holmes, R.L. Buckner, J.W. Smoller, M.R. Sabuncu, Heritability analysis with repeat measurements and its application to resting-state functional connectivity, Proceedings of the National Academy of Sciences (2017) 5521–5526.
[46] T. Ge, M. Reuter, A.M. Winkler, A.J. Holmes, P. HLee, L.S. Tirrell, J.L. Roffman, R.L. Buckner, J.W. Smoller, M.R. Sabuncu, Multidimensional heritability analysis of neuroanatomical shape, Nature Communications 7 (2016) 13291.
[47] T. Ge, T.E. Nichols, D. Ghosh, E.C. Mormino, J.W. Smoller, M.R. Sabuncu, Alzheimer's Disease Neuroimaging Initiative, et al., A kernel machine method for detecting effects of interaction between multidimensional variable sets: an imaging genetics application, Neuroimage 109 (2015) 505–514.
[48] B. Da Mota, V. Fritsch, G. Varoquaux, V. Frouin, J.-B. Poline, B. Thirion, et al., Enhancing the reproducibility of group analysis with randomized brain parcellations, in: MICCAI-16th International Conference on Medical Image Computingand Computer Assisted Intervention-2013, 2013.

[49] H. Huang, J. Yan, F. Nie, J. Huang, W. Cai, A.J. Saykin, L. Shen, A new sparse simplex model for brain anatomical and genetic network analysis, in: Medical Image Computing and Computer-Assisted Intervention—MICCAI 2013, 2013, pp. 625—632.

[50] W. De, Y. Wang, F. Nie, J. Yan, W. Cai, A.J. Saykin, L. Shen, H. Huang, Human connectome module pattern detection using a new multi-graph minmax cut model, in: Medical Image Computing and Computer-Assisted Intervention—MICCAI 2014, Springer International Publishing, 2014, pp. 313—320.

CHAPTER ONE

Multisite Metaanalysis of Image-Wide Genome-Wide Associations With Morphometry

Neda Jahanshad[1], Gennady Roshchupkin[2], Joshua Faskowitz[1], Derrek P. Hibar[1], Boris A. Gutman[1], Hieab H.H. Adams[2], Wiro J. Niessen[2], Meike W. Vernooij[2], M. Arfan Ikram[2], Marcel P. Zwiers[3], Alejandro Arias-Vasquez[3], Barbara Franke[3], Alex Ing[4], Sylvane Desrivieres[4], Gunter Schumann[4], Greig I. de Zubicaray[5], Katie L. McMahon[6], Sarah E. Medland[7], Margaret J. Wright[6], Paul M. Thompson[1]

[1]Keck School of Medicine of USC, Marina del Rey, CA, United States
[2]Erasmus Medical Center, Rotterdam, The Netherlands
[3]Radboud University Medical Center, Nijmegen, The Netherlands
[4]King's College London, London, United Kingdom
[5]Queensland University of Technology (QUT), Brisbane, QLD, Australia
[6]University of Queensland, Brisbane, QLD, Australia
[7]QIMR Berghofer Medical Research Institute, Brisbane, QLD, Australia

Contents

Imaging Genetics
ISBN: 978-0-12-813968-4
http://dx.doi.org/10.1016/B978-0-12-813968-4.00001-8

Abstract

Large-scale distributed analyses of over 30,000 magnetic resonance imaging scans recently detected common genetic variants associated with the volumes of subcortical brain structures. Scaling up these efforts, still greater computational challenges arise in screening the genome for statistical associations at each voxel in the brain, localizing effects using "image-wide genome-wide" testing (voxelwise genome-wide association studies, vGWASs). Here we benefit from distributed computations at multiple sites to metaanalyze genome-wide image-wide data, allowing private genomic data to stay at the site where it was collected. Site-specific tensor-based morphometry is performed with a custom template for each site, using a multichannel registration. A single vGWAS testing 10^7 variants against 2 million voxels can yield hundreds of terabytes (TB) of summary statistics, which would need to be transferred and pooled for metaanalysis. We propose a two-step method, which reduces data transfer for each site to a subset of single-nucleotide polymorphisms and voxels guaranteed to contain all significant hits.

Keywords: Big data; GWAS; Metaanalysis; Multiple comparisons correction; Multisite; Neuroimaging genetics

1. INTRODUCTION

Imaging genetics is an emerging field in which variants in the human genome are related to brain differences, in an attempt to discover specific genetic variants that affect brain development, connectivity, and risk for disease. Genome-wide association studies (GWASs) test for statistical associations between brain measures and over a million single-nucleotide polymorphisms (SNPs), or base-pair variants, in the genome.[1] To simplify the screening effort, studies often focus on one or a handful of measures extracted from brain scans, such as the overall volume of the hippocampus [1]; a recent study of over 30,000 brain magnetic resonance imaging (MRI) scans identified eight genetic loci that were consistently associated with intracranial and subcortical structural volumes, in 50 cohorts worldwide [2]. Given these recent successes with simple volumetric measures, there is great interest in screening the image space more deeply. Each image contains many more features, e.g., at individual voxels, allowing better localization of gene effects and their patterns in the brain. Testing effects of $\sim 10^6$ genetic variants at $\sim 10^6$ voxels requires around 10^{12} statistical tests, but recent volumetric associations achieved significance levels of $P < 10^{-23}$, suggesting that effects can survive even

[1] At each of these SNP locations, there are two possible nucleotides (or alleles), and each individual has two chromosomes that will carry one variant or the other. Therefore, at each SNP location, an individual will have 0, 1, or 2 copies of the minor allele (which is the term used to refer to the least prevalent variant in the population). When testing for statistical associations using an additive genetic model, each SNP is coded as 0, 1, or 2 in each individual.

brute-force Bonferroni correction for millions of GWASs on imaging traits. So far, successful GWASs for single traits have required samples of 10^4-10^6 individuals to discover and independently replicate statistically significant variants [2]. GWAS tests that screen around a million voxels are severely underpowered in individual cohorts because of the large number of tests performed and the stringent statistical criteria needed to establish significance [3,4]. Initial studies show voxelwise genome-wide association studies (vGWASs) are feasible, i.e., image-wide genome-wide testing [4–6], but so far these studies have been underpowered to detect true associations in cohorts of ~1000 individuals. To guard against false positives, it has become standard in genetics to seek replication of results and/or pool data and aggregate evidence across independent cohorts, before associations are considered credible and reproducible. Back then, privacy requirements governing genomic data, and in some cases also brain scans, may prevent raw data from being transferred and shared. This has led to collaborative efforts using protocol harmonization and metaanalysis to aggregate site-specific statistical results. In addition, as datasets become vast and more numerous, there is some benefit to distributing the computation across sites and sending the algorithms to the data rather than centralizing all the data. Therefore, approaches are needed to metaanalyze massive amounts of data from a variety of sources, including image-wide statistics.

Several approaches are required to conduct a distributed voxelwise genome-wide search; first, a registration method is needed to map data from multiple cohorts into a single coordinate space. Without such a registration, voxel measures and statistics would not be comparable across cohorts. Second, dimensionality reduction is commonly applied in neuroimaging studies, but there is no strong prior information on which subsets of the 10^{12} SNP × voxel tests are more likely to support the strongest association signals. Even if the image features were reduced to several thousand voxels, a GWAS evaluating 10^{6}–10^{7} SNPs yields around 400 TB of compressed data to sort and filter at each site, making data transfer of all sites summary statistics to a centralized site, less than ideal.

Although methods for developing optimized single-site vGWAS techniques are also under development [7], our work is applicable to such approaches and focuses on the issues related to the metaanalysis of multisite vGWAS. Here we develop a multisite adaptable protocol using freely available and common neuroimaging software for voxelwise volumetric analysis by tensor-based morphometry (TBM), and we then describe genome-wide association testing in the resulting data. We show its usefulness in a simple univariate approach to vGWAS, though other approaches that perform dimension reduction or multivariate analysis could benefit similarly [8].

Each site conducts statistical analysis on its own cohort locally, yielding results specific to that cohort. We then show how to map each site's results into a common space based on four large and arguably representative cohorts worldwide; we also address the issue of prioritizing data transfer.

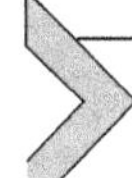

2. METHODS

2.1 Harmonizing Voxelwise Associations for Metaanalysis of Seven Sites

Data from seven separate cohorts were included in this study, listed in Table 1.1.

Table 1.1 Describes the imaging and demographic data from each cohort

	ADNI1	ADNI2	BIG	HCP	Imagen	QTIM	RSS
Scanner	GE, Siemens Philips,	GE	Siemens	GE	GE, Siemens Philips	Siemens Bruker	GE
Field strength	1.5T	3T	1.5 and 3T	3T	3T	4T	1.5T
Location	US multisite	US multisite	Nijmegen, NL	Saint Louis, USA	EU, multisite	Brisbane, AUS	Rotterdam, NL
Voxel size (mm^3)	1.25 × 1.25 × 1.2	1.25 × 1.25 × 1.2	1 × 1 × 1	0.7 × 0.7 × 0.7	1.1 × 1.1 × 1.1	0.9 × 0.9 × 0.9	1 × 1 × 1.6
N	837	815	62	207	80	590	64
Age	75 ± 6.6 (60–89)	72.8 ± 6.6 (48–90)	21.5 ± 1.7 (18–25)	28.7 ± 3.5 (22–35)	14 ± 0.4	22.9 ± 2.8 (18–30)	67.49 ± 11.40

The Queensland Twin Imaging Study (QTIM) and Human Connectome Project (HCP) are family-based studies, but only one person per family was included in this study.
ADNI, Alzheimer's Disease Neuroimaging Initiative; *BIG*, Brain Imaging Genetics; *RSS*, the Rotterdam Study.

Preprocessing: Following the approach in Ref. [2], subcortical and cortical segmentations were performed on 3D anatomical brain MRI scans from each site using FreeSurfer. Quality control protocols were implemented to remove poorly segmented images and outliers, using procedures developed and tested by the Enhancing Neuroimaging Genetics through Meta-Analysis (ENIGMA) Consortium [1,2]. To attempt to reproduce previously reported genetic associations with volumetric measures [2], the integer-valued segmentations of these brain structures were included as part of the fidelity term (the image similarity metric) in a multichannel nonlinear registration approach. The resulting TBM workflow was implemented across all sites to allow voxel-level inferences about genetic associations with regional brain volumetric differences, determined using TBM.

Site-specific and global minimal deformation templates (MDTs): MDTs were constructed using the Advanced Normalization Tools (http://stnava.github.io/ANTs/) software package and accompanying scripts. Approximately 30 scans per cohort were used to create each site-specific template. Images were first linearly aligned to a common space consistent with the Montreal Neurological Institute (MNI) brain template before SyN [9] nonlinear registration was used to obtain deformation fields. To create the multichannel template, a weight was assigned to each channel, corresponding to the contribution of that channel to the total warp. We set the T1-weighted channel to 1, the cortical ribbon to 0.5, and the subcortical segmentations to 0.2. These parameters resulted in stable MDTs; the resulting MDTs remained robust in cases where poor quality scans were deliberately included.

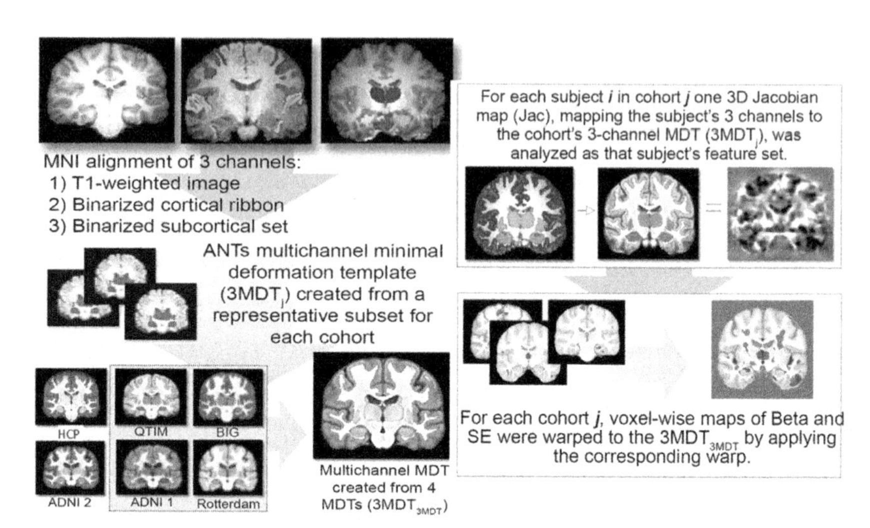

Figure 1.1 Flow diagram of template creation and registration. T1-weighted images run through common software, FreeSurfer, and evaluated to have good-quality cortical, and subcortical parcellations were used along with the FreeSurfer outputs to drive multichannel registrations to a cohort-specific template. The multiple channels were used to reduce variability between cohorts to create a MDT from four datasets. All associations are performed in cohort-specific space, and the transformation from cohort to template space was applied to the resulting statistical maps for metaanalysis. *ADNI*, Alzheimer's Disease Neuroimaging Initiative; *ANT*, Advanced Normalization Tools; *MDT*, minimal deformation template; *QTIM*, The Queensland Twin Imaging Study; *SE*, standard error.

Four of the cohorts (Alzheimer's Disease Neuroimaging Initiative (ADNI-1), the Rotterdam Study, the Queensland study, and Brain Imaging Genetics (BIG); see Fig. 1.1)—representing two older adult and two younger adult cohorts—were used to create a representative MDT for all the cohorts, again using three channels for registration. We chose not to include all sites in the final template but instead use representative sites with varying imaging parameters and demographics; in practice, new sites will often join an ongoing study and continuously reestablishing a template could be impractical.

All cohort-specific MDTs were then registered to the final MDT in the same manner. These warps were maintained for later pooling of statistical maps to a common space. Two alternate methods for template construction and registration were also evaluated: (1) *Single-channel template and registration.* The same T1-weighed images used for multichannel registration were used, but *without* the added FreeSurfer cortical and subcortical labels. (2) *Registration to MNI.* To eliminate the effect of the specific cohorts in the metaanalysis, it might be suggested that an existing template be used for registrations; to this end, we registered all subjects directly to the MNI atlas. This has the advantage of staying consistent even if more subjects or cohorts are added later; however, use of a single template not drawn from the population may introduce other sources of bias in the maps; this also makes single-site level analysis on the extracted det(Jacobian) maps less practical and limits the resulting maps to use in multisite analysis, rather than our proposed approach, which also offers a processing stream for site-specific investigations.

2.2 Voxelwise Associations on Simulated Genetic Data

This work was motivated by distributed "big data" analysis that can accommodate partially private genomic data, as also described in Ref. [8].

As such, we generated a dataset with simulated genetic effects, including 100,000 data points per subject of each dataset to represent an additive genetic effect (0, 1, or 2 at each "genetic locus") using a 2D multinomial distribution with probabilities set to the minor allele frequency (MAF) and 1-MAF. The MAF was uniformly distributed (as approximated from the publicly available ADNI-2 data) and maintained to be greater than 0.1 to avoid rare variants (and by definition <0.5). Files were saved in the .tped format for integration with PLINK software (http://pngu.mgh.harvard.edu/~purcell/plink/).

For each cohort, a univariate GWAS was performed at every individual voxel. To reduce model complexity, covariates including sex and age were removed from the image, and vGWAS was run on residual maps using PLINK2. Data were parallelized across 100 processing nodes, and each PLINK runs using the –mpheno flag over $N_{voxels}/100$ phenotypes. Each voxel output was 8.5 MB in size, generating about 2 TB of data for the downsampled image. As this reduction was by a factor of 4 in each dimension, the full-size image would produce about $4^3 \times 2$ TB of data or ~128 TB.

2.3 Simulating Genetic Data With Predefined Volumetric Effects

To simulate SNPs with localized associations with specific regions of the brain, we set a handful of SNPs to have marginal effects with measures from all cohorts. From the imaging data, regional volumetric measures were extracted and defined according to ENIGMA protocols. Of the generated SNPs, five were designed to meet certain volumetric summary criteria for each individual cohort.

1. This SNP with MAF = 0.1 was simulated to be marginally ($z = 1.96$) associated with average bilateral thalamic volume (after removing the intracranial volume (ICV) effect).
2. Same as #1 but with MAF = 0.3.
3. SNP with MAF = 0.1 was generated to have an effect size of $z = \min(N/10, 5)$ when regressed against

bilateral hippocampal volume (HV), after the ICV effect is removed, such that the significance was related to the cohort size, N, yet was not excessive ($|z| < 5$).

4. Same as #3 but with MAF = 0.3.
5. An SNP with MAF = 0.3 was set to be similarly associated with ICV, a feature whose effect we intended to remove from the voxelwise associations, as TBM is performed on images that have been linearly aligned to include scaling after skull stripping, and therefore the effect of ICV was not included here.

To enforce these associations, a correlation coefficient was determined from the set Z-statistic ($z = 1.96$ for case 1, 2 above). As vectors with mean 0 have corr = cos(*theta*), where *theta* is the angle between them, we centered and orthogonalized the response variable (e.g., HV) with a QR-decomposition and scaled back; this method does not lead to the integer values 0, 1, and 2 needed, so the values were rounded and correlation values were recomputed, and the process was iterated (simulating new SNP values) until the final correlation between the simulated SNP value and the true volume of interest was within ± 0.1 of the desired value.

As this work is intended to be a proof of concept, we downsampled the images by a factor of 4 for the genome-wide analysis, so that the images contained ~30,000 voxels (this varied slightly by site). The final MDT had 31,725 voxels.

2.4 Reduction of Data Transfer to Eliminate Negative Exchange

The inverse-variance weighted metaanalytical P-value for SNP i at trait (voxel) v, is [10]: $p_{MA}(i, v) = 2\Phi(|-Z_{MA}(i, v)|)$, where Φ represents the normal transformation, and the Z-score is the Beta value of the association (or the unnormalized effect size) divided by the standard error from the regression performed in the GWAS.

The metaanalyzed Z-score is then given by the overall Beta divided by the overall standard error (SE). Using the inverse-variance weighted metaanalysis formula described by Willer et al. [10], we get:

$$Z_{MA}(i,v) = \frac{\beta(i,v)}{SE(i,v)} = \frac{\sum_j \beta_j(i,v) \times SE_j^{-2}(i,v) \Big/ \sum_j SE_j^{-2}(i,v)}{\sqrt{1 \Big/ \sum_j SE_j^{-2}(i,v)}} \tag{1.1}$$

here $\beta_j(i, v)$ is site j's effect size, and $SE_j(i, v)$, the site's standard error for SNP i and trait v. Statistical significance implies that the P-value is less (more extreme) than a given threshold (p_{cutoff}), or similarly, the magnitude of the statistic, denoted by Z, must be greater than a specific threshold ($|Z_{cutoff}|$):

$$|Z_{MA}(i,v)| \geq 0.5\Phi^{-1} p_{cutoff}(i,v) = |Z_{cutoff}(i,v)| \tag{1.2}$$

If an SNP (i_k) at a given v_k exceeds this statistical significance threshold, then the maximum Z-score for that SNP across all voxels v must also pass the threshold. In other words, if the SNP is image-wide genome-wide significant at a particular voxel (v_T), then the Beta and SE of that SNP at that voxel $\beta(i_k, v_T)$, $SE(i_k, v_T)$ will lead to a Z-score less than or equal to the Z obtained from the maximal Beta and the minimum *SE* in voxels contributing to a collapsed region.

To formulate this:

$$\underset{v}{\operatorname{argmax}}(Z_j(i_k, v)) \equiv \frac{\underset{v}{\operatorname{argmax}}(\beta_j(i_k, v_k))}{\underset{v}{\operatorname{argmin}}(SE_j(i_k, v_k))}$$

$$= \frac{\beta_j(i_k, v_{k1}) \times SE_j^{-2}(i_k, v_{k2}) / SE_j^{-2}(i_k, v_{k2})}{\sqrt{SE_j^2(i_k, v_{k2})}}$$

$$\geq \frac{\beta_j(i_k, v_T) \times SE_j^{-2}(i_k, v_T) / SE_j^{-2}(i_k, v_T)}{\sqrt{SE_j^2(i_k, v_T)}} \tag{1.3}$$

Here, the maximal statistic for the Beta and the minimal statistic for the *SE* may come from different voxels, so we denote them as $k1$ and $k2$. The right hand side of the equation is Eq. (1.1), with the *Z*-score set for SNP (i_k) at the voxel where the significance threshold was reached (v_T). It will always be less than or equal to the *Z*-score obtained from the maximal Beta and minimal *SE* for a set of voxels.

Therefore, we can collapse the image localization information to take only the most extreme values for the SNP across the full image. We note that in accordance with the inverse-variance weighted metaanalysis formula, to ensure the maximal statistic for each site, we take the *Z*-statistics to be the most extreme Beta divided by the standard error. If the most extreme +*Z*-scores for each SNP are taken across all sites, then it can only exceed the significant *Z*-score at i_k, v_k.

$$Z_{MA}(i_k) = \frac{\sum_j \underset{v}{\operatorname{argmin}}\left(\beta_j(i_k, v)\right) \times \underset{v}{\operatorname{argmin}}\left(SE_j(i_k, v)\right)^{-2}}{\sqrt{1 \Big/ \sum_j \underset{v}{\operatorname{argmin}}\left(SE_j(i_k, v)\right)^{-2}}} \geq Z_{MA}(i_k, v_T) \tag{1.4}$$

Note positive and negative Z's are possible, so the same must be considered for $-Z$, or argmin(Beta), taking into account the sign.

Only SNPs with a metaanalyzed statistic $|Z_{MA}(i_k)| \geq Z_{cutoff}$ will be subjected to full metaanalysis across all voxels.

Voxelwise metaanalysis would tend to filter out localized false positives; for example, if one SNP shows a false positive effect in one voxel in one cohort, it is still unlikely to also have a false effect at the same voxel in a different cohort, so the metaanalysis would not necessarily show a significant effect. However, when collapsing the image the lack of localization can enhance false positives. Given the large number of voxels present in an image (1,869,764 in the full-resolution image and 31,725 in the downsampled) using this TBM method, the probability that an SNP will reach the threshold for significance at **any** voxel is high. Therefore, we also demonstrate the effect of this approach in segmented images and show the effect it has on the reducing the data required for transfer.

The procedure described here involves the following site-specific steps (for which harmonized scripts would be provided):

1. Creating the cohort-specific template and defining its mapping to the overall template (this can be done at

the central site, or the template and its mapping to the overall template is sent to the central site).

2. Proposed 3-channel registration of all subjects in cohort to the cohort-specific MDT for TBM analysis.
3. Voxelwise GWAS at the site level.
4. Finding the minimum and maximum statistics (Beta and *SE*) across all voxels (in the full image or a given parcellation) for each SNP and sending this information to the central site.

Regions of interest (ROIs) are delineated on the overall-MDT by the central site, and inverse warps from the cohort MDT are applied to the labels using a nearest neighbor interpolation. Each cohort can then perform voxelwise GWAS in the cohort's own template space and transfer (back to the central site) the required information for the regions.

The data transfer is performed in two steps (illustrated in Fig. 1.2):

1. Sending these minimum and maximum results ($\pm Z$).
   a. As data are provided for each SNP, this equates to sending two full GWAS results to the central site for each parcellation of the image (for example, for one hemisphere).
   b. At this stage—the central site pools all sites' results, and from the resulting metaanalysis determines which SNPs would not reach genome-wide significance; these SNPs are then removed from the second stage of data transfer. The central site then determines what SNP set is needed for each different voxel in the parcellations for further refinement of the data and localization of the associations.
2. Sending full localized results for only the reduced set of SNPs for each chosen parcellation of the brain (whole brain or ROIs).

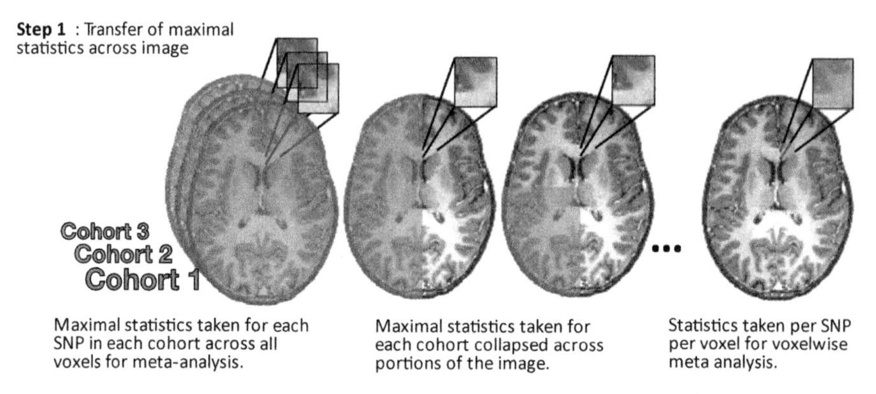

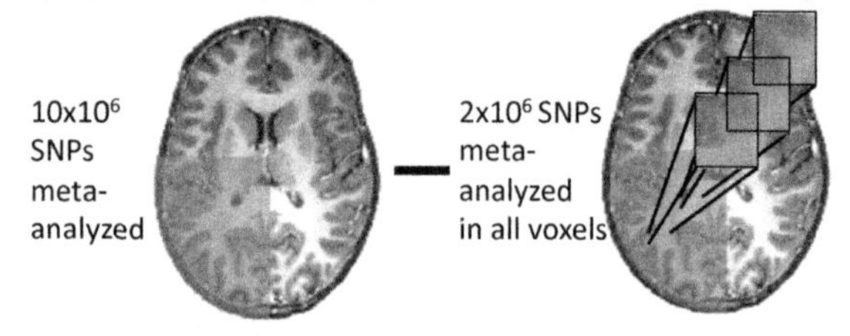

Figure 1.2 The maximal statistics (both positive and negative *Z*-statistics) for each single-nucleotide polymorphism (SNP) were taken across all statistical tests conducted in the collapsed regions for each cohort and sent to the central site for metaanalysis. These maximal statistics were then metaanalyzed across cohorts, where only a fraction of SNPs in certain partitions are image-wide genome-wide significant. In Step 2, finer, voxel-level statistics are then only transferred for SNPs meeting the significance criterion in the collapsed regions from Step 1, avoiding terabytes of data transfer and analysis from SNPs and voxels not reaching significance levels. Various ways of parcellating the voxels in the image are shown. Collapsing across all voxels already leads to a 16% reduction in data transfer.

a. Note it is possible to further break down a given parcellation and repeat step one in refined regions, but for simplicity, we assume this step is performed only once.

b. This would be what would be needed to transfer for a standard metaanalysis, but now the SNP set and potentially the voxel set, is greatly reduced.

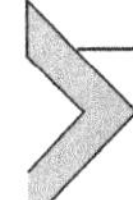

3. RESULTS

3.1 Simulated Associations of Fixed Genetic Effects

Fig. 1.3A shows the effect of a single variant with deliberately prescribed effects on thalamic volume (SNP 1 above) mapped using multiple possible voxelwise analysis methods. We used a multichannel approach where the cortical and subcortical volume segmentations were used as added registration channels to help drive the registration (MDT creation and intersubject registration). This multichannel approach showed visibly greater specificity in detecting the thalamic signal, perhaps because of better registration accuracy, and the benefit of accurately registering the thalamic boundaries across subjects and cohorts. Fig. 1.3B shows the effect of the multichannel registration approach when metaanalyzing a fixed SNP effect (SNP #4 above) on a voxelwise level. Voxel-level analysis also showed regional specificity, giving FDR-significant voxels bilaterally in the hippocampus.

3.2 Data Reduction by Transferring a Reduced Set of Single-Nucleotide Polymorphisms

When analyzing the brain in full, and collapsing the image to take the most extreme positive and negative statistics across all voxels for metaanalysis, we find, as may be expected from the multiple comparisons, that this did, though not greatly, reduce the number of SNPs that could potentially survive multiple comparisons correction after metaanalysis.

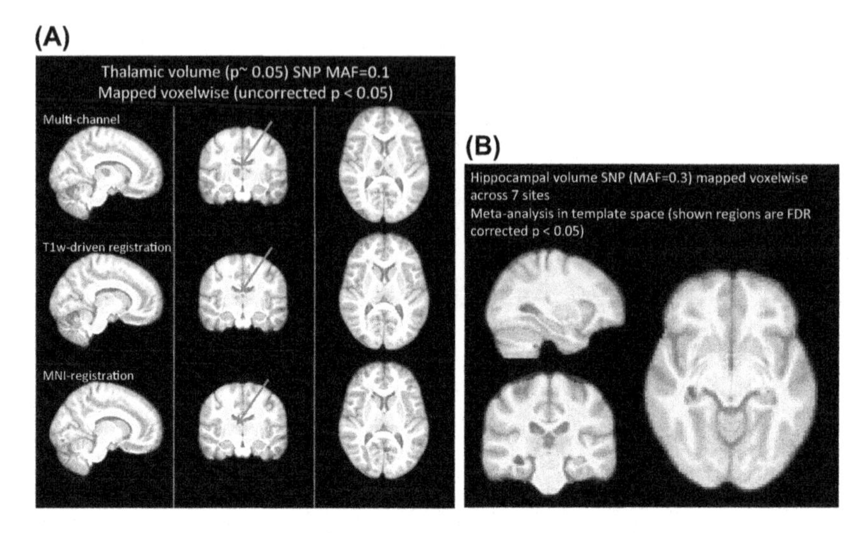

Figure 1.3 (A) An single-nucleotide polymorphisms (SNP) with MAF = 0.1 was simulated to be marginally ($z = 1.96$) associated with average bilateral thalamic volume in a single cohort (after removing intracranial volume). The effect of maintaining specificity to the thalami was compared between multiple templates. No method produced voxelwise significant maps; however, evaluating the uncorrected association results of the methods shows greater thalamic effects in the multichannel method. (B) An SNP with MAF = 0.3 was generated for each of seven cohorts, to have $z = \min(N/10, 5)$ such that the significance was related to cohort size, yet was not excessive ($|z| < 5$). Beta and standard error maps for all cohorts were mapped to template space, and voxelwise metaanalysis revealed associations localized to both hippocampi. *FDR*, false discovery rate; *MAF*, minor allele frequency; *MNI*, Montreal Neurological Institute.

With a strict Bonferroni correction accounting for all SNPs (100,000) and voxels (31,725), 84% of SNPs (or 83,954 of the 100,000) would be needed, already accounting for an approximate 16% reduction in data transfer.

Dividing the image into two—the left and right hemispheres—led to an overall reduction to 82,495 SNPs across the brain. However, data transfer could be further reduced as only 78,041 were found in one hemisphere, so for 4454 of the 82,495 SNPs, data for ½ the image would not need to be transferred.

Further breaking down the image into bilateral ROIs, including bilateral subcortical regions, such as hippocampus, putamen, etc., as well as cortical parcellations, including anterior, posterior temporal/parietal lobes, cingulate gyrus etc., and remaining unlabeled sections, resulted in an even further reduced dataset for transfer. Though a total of 82,711 SNPs were still identified, SNPs were retained for small ROIs. Most of the smaller ROIs were less than 1000 voxels in size, and when their voxels were collapsed, they held between 0 and a few hundred possibly significant SNPs out of the full 100,000. The ROI with the greatest amount of potential significant SNPs (51% of the total) for follow-up included the bilateral superior frontal gyrus, which made up about 9% of the image.

Separating the abovementioned regions into their respective L and R hemispheres (84 total, not listed here for brevity) resulted in a drastically reduced SNP set of only 73% of the total (73,250 of the 100,000) for the full image. However, once again, as certain SNPs were only significant in certain regions, the transfer of their information from nonsignificant voxels is not necessary, further reducing the data transfer to less than half of the data generated. The superior frontal gyrus again held the most number of possible SNPs, with 18,528 in the left hemisphere and 22,389 in the right. Clearly separating this region into L and R already reduced the total number of SNPs by ~10% of the total, and 20% of those identified for the region itself. Fig. 1.4 plots the number of possibly significant

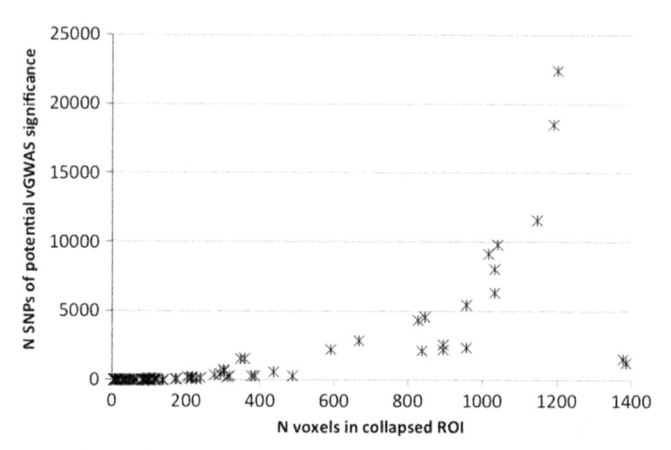

Figure 1.4 Plot showing the number of potentially significant single-nucleotide polymorphisms (SNPs) metaanalyzed in collapsed regions of interest (ROIs) of a given size. The ~1400 voxel ROIs in the cerebellar region do not follow a similar trend to that of other brain regions. vGWAS, voxelwise genome-wide association studies.

SNPs as a function of ROI size when the ROI is collapsed to the most extreme statistic. Clearly as the number of ROIs increases, the data transfer in the first step also increases, but numbers are orders of magnitude less than if the images were transferred on a voxel by voxel level.

4. DISCUSSION

Here we showed how to extend large-scale metaanalytic genetic association studies to image-wide analyses, by including steps to pool data across templates and make intersite transfer of data more efficient. A distributed parallel computation can be highly beneficial as cohorts increase in size and add to a study or as new cohorts join an analysis. Because of the high levels of computation, it is also not practical to rerun analyses at every site, so our approach makes use of common analyses that many sites have already performed on structural MRI scans to make volume measurements, leading to a harmonized protocol for voxelwise association studies. Although prior voxel-level metaanalyses have been performed, they involve pooling data from published results, which may highlight the association only at specific points in particular populations, and the mapping between regions and the summary statistics is unknown [11]. Using simulated genetic markers, we show our technique can maintain full structure-level volume associations (i.e., effects on HV) when mapping out voxel-level associations, not only in a single cohort but also when data are pooled across a diverse set of cohorts.

In our analysis, we generated 100,000 data points to represent genomic markers with varying allelic frequencies. Here, all SNPs were generated independently of the others; therefore we are not evaluating any effect of linkage seen in standard GWAS. An exhaustive search of all SNPs and voxels may be avoidable by using dimensionality reduction methods to both the image and genome. Reduced rank regression, parallel independent component analysis, canonical covariates analysis, and nonlinear machine learning approaches have all been proposed to fit sets of genetic predictors to imaging data. SNP set selection methods, annotation methods, and Bayesian priors have all been proposed to prioritize sets of SNPs in models rather than search them in a bias-free way. Other methods are also possible, though they may not guarantee all significant SNPs will be captured [8,12,13].

However, at the time of writing, reproducible associations have been hard to identify with these methods, whereas volumetric associations have been reproduced across 50 cohorts using univariate association testing [2]. Fortunately, the inefficiencies in these standard univariate GWAS approaches will be overcome in the future.

Our work here shows, however, a method applicable to any voxelwise or vertexwise GWAS that is to be extended to a metaanalysis. Rather than transferring all data, a multistage process can be conducted to break down the image into ROIs, find the most significant measures for each SNP at those locations and determine whether they would reach global significance in the metaanalysis. Only after the first screening more information would be requested about specific SNPs of interest. This allows a significant reduction in data transfer between sites as well as expedited metaanalysis that would not need to read hundreds of TB of text files and store them in memory simultaneously.

REFERENCES

[1] J.L. Stein, S.E. Medland, A.A. Vasquez, D.P. Hibar, R.E. Senstad, A.M. Winkler, et al., Identification of common variants associated with human hippocampal and intracranial volumes, Nature Genetics 44 (May 2012) 552–561.

[2] D.P. Hibar, J.L. Stein, M.E. Renteria, A. Arias-Vasquez, S. Desrivieres, N. Jahanshad, et al., Common genetic variants influence human subcortical brain structures, Nature 520 (April 9, 2015) 224–229.

[3] S.E. Medland, N. Jahanshad, B.M. Neale, P.M. Thompson, Whole-genome analyses of whole-brain data: working within an expanded search space, Nature Neuroscience 17 (June 2014) 791–800.

[4] J.L. Stein, X. Hua, S. Lee, A.J. Ho, A.D. Leow, A.W. Toga, et al., Voxelwise genome-wide association study (vGWAS), Neuroimage 53 (November 15, 2010) 1160–1174.

[5] D.P. Hibar, J.L. Stein, O. Kohannim, N. Jahanshad, A.J. Saykin, L. Shen, et al., Voxelwise gene-wide association study (vGeneWAS): multivariate gene-based association testing in 731 elderly subjects, Neuroimage 56 (June 15, 2011) 1875–1891.

[6] M. Vounou, E. Janousova, R. Wolz, J.L. Stein, P.M. Thompson, D. Rueckert, et al., Sparse reduced-rank regression detects genetic associations with voxel-wise longitudinal phenotypes in Alzheimer's disease, Neuroimage 60 (March 2012) 700–716.
[7] M. Huang, T. Nichols, C. Huang, Y. Yu, Z. Lu, R.C. Knickmeyer, et al., FVGWAS: fast voxelwise genome wide association analysis of large-scale imaging genetic data, Neuroimage 118 (September 2015) 613–627.
[8] Q. Li, T. Yang, L. Zhan, D. Hibar, N. Jahanshad, J. Ye, et al., Large-scale collaborative genetic studies of risk SNPs for Alzheimer's disease across multiple institutions, in: Presented at the International Symposium for Biomedical Imaging – ISBI, Prague, Czech Republic, 2016 (submitted).
[9] B.B. Avants, C.L. Epstein, M. Grossman, J.C. Gee, Symmetric diffeomorphic image registration with cross-correlation: evaluating automated labeling of elderly and neurodegenerative brain, Medical Image Analysis 12 (February 2008) 26–41.
[10] C.J. Willer, Y. Li, G.R. Abecasis, METAL: fast and efficient meta-analysis of genomewide association scans, Bioinformatics 26 (September 1, 2010) 2190–2191.
[11] G. Salimi-Khorshidi, S.M. Smith, J.R. Keltner, T.D. Wager, T.E. Nichols, Meta-analysis of neuroimaging data: a comparison of image-based and coordinate-based pooling of studies, Neuroimage 45 (April 15, 2009) 810–823.
[12] Y. Li, J. Wang, T. Yang, J. Chen, L. Lui, L. Zhan, et al., Identification of Alzheimer's disease risk factors by Tree-Structured Group LASSO screening, in: Presented at the International Symposium for Biomedical Imaging – ISBI. Prague, Czech Republic, 2016 (submitted).
[13] M. Lorenzi, B. Gutman, D. Hibar, A. Altmann, N. Jahanshad, P. Thompson, et al., Validating partial least squares modelling for imaging-genetics in Alzheimer's disease, in: Presented at the International Symposium for Biomedical Imaging – ISBI, Prague, Czech Republic, 2016 (submitted).

CHAPTER TWO

Genetic Connectivity—Correlated Genetic Control of Cortical Thickness, Brain Volume, and White Matter

Daniel A. Rinker[1], Neda Jahanshad[1], Derrek P. Hibar[1], Joshua Faskowitz[1], Katie L. McMahon[2], Greig I. de Zubicaray[3], Margaret J. Wright[2], Paul M. Thompson[1]
[1]Keck School of Medicine of USC, Marina del Rey, CA, United States
[2]University of Queensland, Brisbane, QLD, Australia
[3]Queensland University of Technology (QUT), Brisbane, QLD, Australia

Contents

Imaging Genetics
ISBN: 978-0-12-813968-4
http://dx.doi.org/10.1016/B978-0-12-813968-4.00002-X

Abstract

Magnetic resonance imaging (MRI) and diffusion tensor imaging (DTI) measures of brain volume, cortical thickness, and white matter (WM) integrity are commonly used in imaging genetics studies, but the genetic relationship between these measures is not well understood. Here we use structural equation models in a twin design to test the genetic correlation between these common imaging measures. MRI and DTI data from 442 participants (mean age: 23.5 years ±2.1 SD; 151 women; 98 monozygotic pairs, 123 dizygotic pairs) were analyzed using standardized Enhancing Neuroimaging Genetics through Meta-Analysis protocols. We found significant genetic associations between measure of the integrity (fractional anisotropy, or FA) of several WM tracts and subcortical volume regions of interest, notably the thalamus and pallidum. Correlation was low between cortical thickness or volume and DTI measures from the WM. Total cortical surface area was, however, highly correlated with FA in several WM regions and all of the subcortical volume regions. These results may be useful for future studies assessing specific genetic associations and offer insight into the genetics underlying common imaging measures.

Keywords: Brain volume; Cortical thickness; Diffusion imaging; Genetic correlation; Imaging genetics; Structural MRI; White matter integrity

1. AIMS

Understanding the degree to which different brain measures are affected by shared genetic influences is valuable in explaining brain changes during development and emerging pathology; it can also help experimental design in the field of imaging genetics [1]. Brain structures are under strong genetic control [2], but patterns of common genetic influence across multiple brain measures (and across different imaging modalities) are still largely unknown.

Understanding the genetic relationship between different brain measures may offer biologically meaningful targets for genetic analysis, revealing how they are interconnected. In searching for genes that influence brain measures, it would be advantageous to select imaging measures that are genetically distinctive or unique—discovering genes that help shape multiple brain regions is also of great interest.

Twin and family studies can estimate the degree of common genetic influence underlying any two traits (called the genetic correlation or r_g). By clustering cortical regions that are genetically correlated, Fjell et al. recently noted modules or sectors of the cortex that appear to develop and age in distinctive ways [3,4]. Panizzon et al. found that cortical surface area and thickness are influenced by separate genetic factors—they suggested that surface or thickness measures may be better targets for gene discovery studies than cortical volume measures, which contain genetic and phenotypic aspects of both [5]. A corollary of this work is to examine measure across modalities and possibly to develop a transmodality measure for gene discovery.

Diffusion tensor imaging (DTI) is commonly acquired along with standard T1-weighted structural magnetic resonance imaging (MRI) scans and has the potential to describe white matter (WM) integrity across the entire brain. The relationship between DTI metrics and brain morphometry is largely unstudied. WM tracts imaged in DTI interconnect many subcortical regions commonly studied in imaging genetics. They also develop in concert with cortical and subcortical gray matter as neural pruning and fiber organization occurs [6]. The genetic control of this development is not yet understood.

Here we examined 442 healthy, young adult twins to estimate the genetic correlation between measures from 109 brain regions of interest (ROIs; 21 measures of WM fractional anisotropy (FA), 15 of subcortical volume, 72 of cortical thickness, and the intracranial volume). These measures were computed with standardized protocols developed by the Enhancing Neuroimaging Genetics through Meta-Analysis (ENIGMA) Consortium. The ENIGMA Consortium is a large multisite imaging genetics collaboration involving over 185 laboratories across 35 countries. ENIGMA recently published a genome-wide association study of subcortical volumes, using MRI and genotyping data from 30,717 individuals [7]; similar studies of DTI and cortical measures are now underway.

Here we evaluated the genetic relationships among WM, subcortical volumetrics, and cortical thickness and surface area. Given the developmental and physical links between WM and cortical thickness, surface area, and subcortical volumetrics, we hypothesized that the underlying genetic determinants of each of those measures would show some overlap with those measuring WM integrity.

2. METHODS

Bivariate genetic correlations were computed between ROIs for three imaging measures: subcortical volume, cortical thickness, and DTI FA, as described in the ENIGMA protocols.

2.1 Subject Information

A total of 442 subjects (mean age: 23.5 years ±2.1 SD; 151 women; 98 monozygotic (MZ) pairs, 123 dizygotic (DZ) pairs) were included; all subjects underwent structural T1-weighted brain MRI and DTI scans. All subjects were of European ancestry from 221 families. Subjects were recruited as part of a large-scale 5-year twin study examining healthy young adult Australians using structural and functional MRI and DTI [8].

2.2 Image Acquisition

Structural and diffusion-weighted (DW) whole-brain MRI scans were acquired for every participant (on a 4T Bruker Medspec scanner). T1-weighted images were acquired with an inversion recovery rapid gradient echo sequence (TI/TR/TE = 700/1500/3.35 ms; flip angle = 8 degrees; slice thickness = 0.9 mm, with a 256^3 acquisition matrix).

DW images were acquired using single-shot echo planar imaging with a twice-refocused spin-echo sequence to reduce eddy-current-induced distortions. A 3-min, 30-gradient acquisition was designed to optimize signal-to-noise ratio for diffusion tensor estimation (58). Imaging parameters were: TR/TE = 6090/91.7 ms, FOV = 23 cm, with a 128 × 128 acquisition matrix. Each 3D volume consisted of 55 2-mm thick axial slices and $1.8 \times 1.8\ \mathrm{mm}^2$ in-plane resolution. One hundred and five images were acquired per subject: 11 with no diffusion sensitization (i.e., T2-weighted b_0 images) and 94 DW images ($b = 1149\ \mathrm{s/mm}^2$) with gradient directions uniformly distributed on the hemisphere.

2.3 Image Preprocessing

All images were processed as described by the publicly available ENIGMA image analysis protocols (http://enigma.ini.usc.edu/protocols/imaging-protocols/).

2.4 Establishing Zygosity, Genotyping, and Imputation

Standard polymerase chain reaction (PCR) methods and genotyping were used to establish zygosity objectively by typing nine independent DNA microsatellite polymorphisms (polymorphism information content >0.7).

Blood group (ABO, MNS, and Rh) was used to verify results along with phenotypic data (hair, skin, and eye color), providing a probability of accurate zygosity classification >99.99%. Standard manufacturer protocols were used on the Human610-Quad BeadChip (Illumina) to analyze genomic DNA samples (Infinium HD Assay; Super Protocol Guide; Rev. A, May 2008). Genotypes were imputed by mapping to HapMap (Release 22, Build 36) with MACH (http://www.sph.umich.edu/csg/abecasis/MACH/index.html).

2.5 Cross-Twin Cross-Trait Analysis

To identify common genetic or environmental factors modulating cortical thickness, subcortical volume, and DTI FA measures, we used a "cross-twin cross-trait" analysis [9]. Covariance matrices for the MRI and DTI measures were computed between the MZ twins who share all the same genes, and the DZ twins who on average share half of their genetic polymorphisms.

Using OpenMx software (http://openmx.psyc.virginia.edu/), covariance matrices were entered into a multivariate structural equation model (SEM) to estimate the relative contributions of additive genetic (A), shared environmental (C), and unique environmental (E) components to the population variances and covariances of the observed variables. The E component also contains experimental measurement error and is assumed to be independent between both twins in a pair.

In multivariate SEM, it is expected that there are common genetic and environmental factors affecting various brain measures. We can estimate the variance of the common genetic and environmental components from the total population variance by calculating the difference in the covariances between the MZ and DZ twins within the same individual (cross-trait within individual) and also between one phenotype in one twin with the other phenotype in the second twin (cross-twin cross-trait). With this multivariate SEM, we obtain r_a and r_c, which denote the additive genetic and shared environmental influences on the correlations between the two phenotypes, respectively.

The cross-trait within individual correlation (the correlation between two brain measures, FA and volume for example, in twin 1 or in twin 2) can be split into the additive genetic and shared and unique environmental components (e.g., $A_{V,i}$, $C_{V,i}$, and $E_{V,i}$ for each ROI value), and the correlation coefficients between $A_{V,i}$ and $A_{FA,i}$, $C_{V,i}$ and $C_{FA,i}$, and $E_{V,i}$ and $E_{FA,i}$, are indicated by r_a, r_c, and r_e, respectively. The cross-trait cross-twin correlation is shown as $A_{V,i}$ and $A_{FA,j}$, and $C_{V,i}$ and $C_{FA,j}$ for the FA value in twin i and the volume value in twin j, where $i, j = 1$ or 2 and $i \neq j$. Because the unique environmental factors between subjects are considered to be independent, there is no r_e term for $E_{V,i}$ and $E_{T,j}$.

Using the path diagram, we derive the covariance across the two phenotypes within the same subject (or separately in the two subjects) by multiplication of the path coefficients for the closed paths (Fig. 2.1).

For example, covariance between the FA values in *twin 1* and the volume in *twin 2* is equal to $a_V \cdot r_a \cdot a_T + c_V \cdot r_c \cdot c_T$ for MZ twins and $a_V \cdot ½r_a \cdot a_T + c_V \cdot r_c \cdot c_T$ for DZ twins. Paths connecting the same phenotype are identical to a univariate SEM model [10]. MZ twins have a correlation coefficient of

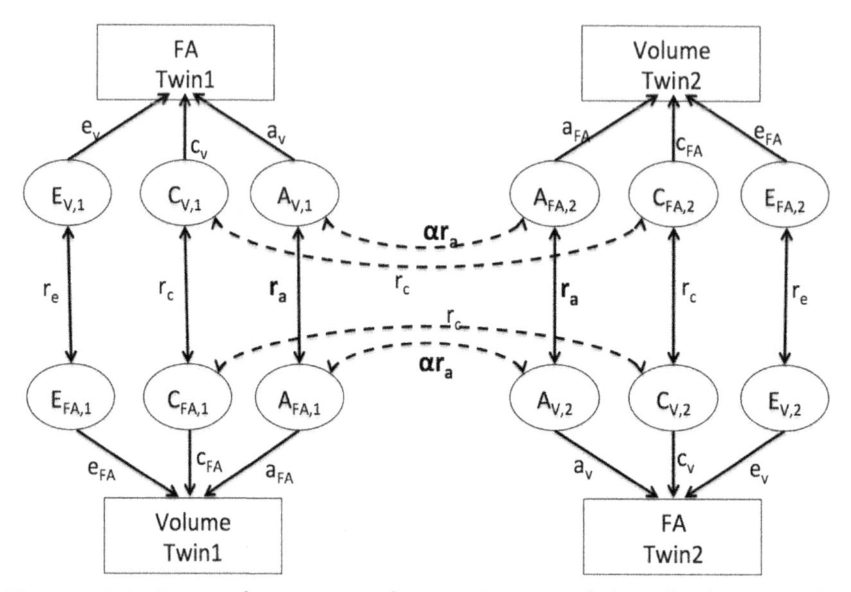

Figure 2.1 Example structural equation model path diagram for bivariate association. *FA*, fractional anisotropy.

1 for A1 and A2, whereas DZ twins have 0.5. By definition for the shared environment, C1 and C2 is always a correlation coefficient of 1. E1 and E2 are assumed to have no correlation.

It is common in twin studies to test whether the observed measures are best modeled using a combination of additive genetic, shared, and unshared environmental factors, or whether only one or two of these factors is sufficient to explain the observed pattern of intertwin correlations. More details on model selection are described in Jahanshad et al. [11]. Here we used the full set of path coefficients in each test as they achieved significance.

2.6 Phenotypic Correlations

To assess the phenotypic relationship across measures, correlation matrices were generated using `cor.test` as a part of the R software package (https://www.r-project.org/), across all measurements entered into the genetic analysis.

2.7 Multiple Comparisons Correction

To control for multiple comparisons, the standard Benjamini and Hochberg [12] false discovery rate (FDR) procedure was employed in all statistical tests: in determining the best overall model for the SEM cross-twin cross-trait analysis and in the phenotypic and genetic correlation analysis ($q = 0.05$).

3. RESULTS

Region abbreviations may be found in the Glossary section. Bivariate genetic correlations between subcortical volume and FA measures are shown in Fig. 2.2.

For tests of genetic correlation between cortical thickness and FA, the only significant correlations were of left and right hemisphere surface area, intracranial volume (ICV), and various DTI tracts listed below in Fig. 2.3.

Similarly in tests of genetic correlation between cortical thickness and volume, only surface area and ICV were significant, listed in Fig. 2.4.

The results of the phenotypic correlations between FA and volume are shown in Fig. 2.5.

Phenotypic correlations between cortical thickness and volume are shown in Fig. 2.6.

In the phenotypic correlation test between thickness and FA, the only significant findings were between the *genu* of the corpus callosum (GCC) and left surface area, right surface area, and ICV ($r = 0.22$, 0.23, 0.22, respectively, $q < 0.05$) and the anterior *corona radiata* and left surface area and right surface area ($r = 0.23$, 0.23, respectively, $q < 0.05$).

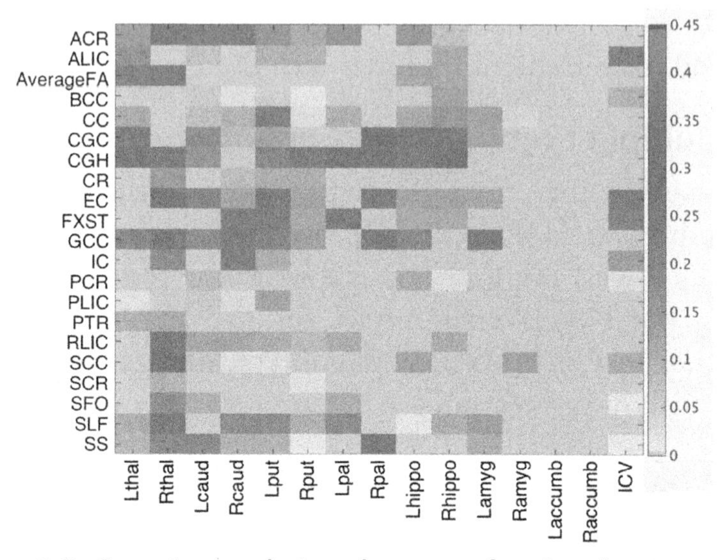

Figure 2.2 Genetic correlations between fractional anisotropy in diffusion tensor imaging (DTI) white matter tracts and subcortical volume regions of interest. Colored elements (light gray in print versions) indicate significant associations after false discovery rate correction. Warmer colors (darker colors in print versions) indicate higher genetic correlation (r_g) values. Some very small structures (e.g., the nucleus accumbens) did not show significant correlations with any of the DTI measures. *Accumb*, nucleus accumbens; *ACR*, anterior corona radiata; *ALIC*, anterior limb of the internal capsule; *Amyg*, amygdala; *BCC*, body of the corpus callosum; *Caud*, caudate; *CC*, corpus callosum; *CGC*, cingulum; *CHG*, hippocampal part of the cingulum; *CR*, corona radiata; *EC*, external capsule; *FXST*, fornix/stria terminalis; *GCC*, genu of the corpus callosum; *Hippo*, hippocampus; *IC*, internal capsule; *ICV*, intracranial volume; *Pal*, globus pallidus; *PCR*, posterior corona radiata; *PLIC*, posterior limb of the internal capsule; *PTR*, posterior thalamic radiation; *Put*, putamen; *RLIC*, retrolenticular part of the internal capsule; *SCC*, subcallosal cingulate white matter; *SCR*, superior region of corona radiata; *SFO*, superior fronto-occipital fasciculus; *SLF*, superior longitudinal fasciculus; *SS*, sagittal striatum; *Thal*, thalamus.

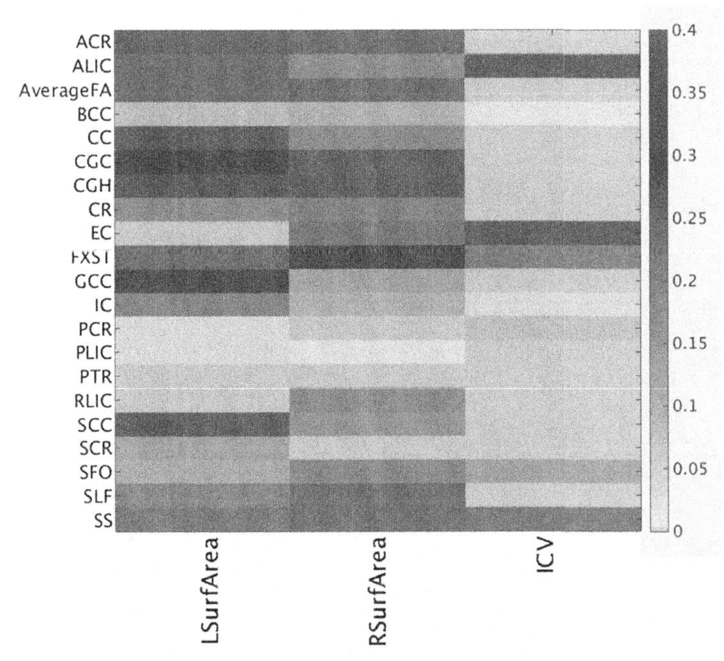

Figure 2.3 Genetic correlations between fractional anisotropy in diffusion tensor imaging white matter tracts, cortical surface area, and intracranial volume (ICV). Colored elements (light gray in print versions) indicate significant associations after false discovery rate correction. Warmer colors (darker colors in print versions) indicate higher genetic correlation (r_g) values. *ACR*, anterior corona radiata; *ALIC*, anterior limb of the internal capsule; *BCC*, body of the corpus callosum; *CC*, corpus callosum; *CGC*, cingulum; *CHG*, hippocampal part of the cingulum; *CR*, corona radiata; *EC*, external capsule; *FXST*, fornix/stria terminalis; *GCC*, genu of the corpus callosum; *IC*, internal capsule; *PCR*, posterior corona radiata; *PLIC*, posterior limb of the internal capsule; *PTR*, posterior thalamic radiation; *RLIC*, retrolenticular part of the internal capsule; *SCC*, subcallosal cingulate white matter; *SCR*, superior region of corona radiata; *SFO*, superior fronto-occipital fasciculus; *SLF*, superior longitudinal fasciculus; *SS*, sagittal striatum; *SurfArea*, cortical surface area.

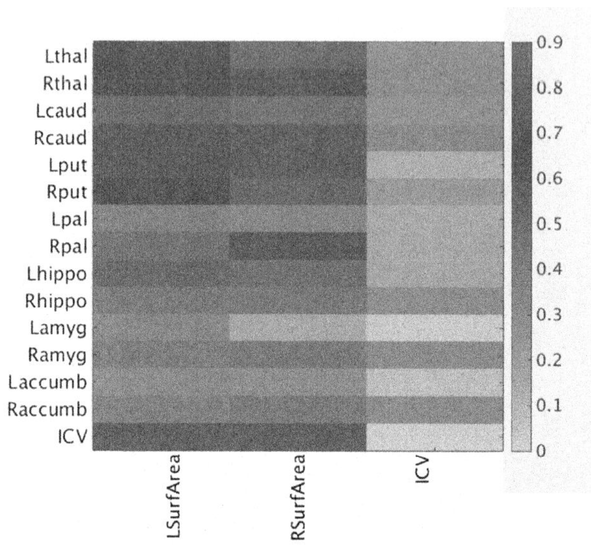

Figure 2.4 Genetic correlations between subcortical volume regions of interest and cortical surface area measures. Colored elements (light gray in print versions) indicate significant associations after false discovery rate correction. Warmer colors (darker colors in print versions) indicate higher genetic correlation (r_g) values. *Accumb*, nucleus accumbens; *Amyg*, amygdala; *Caud*, caudate; *Hippo*, hippocampus; *ICV*, intracranial volume; *Pal*, globus pallidus; *Put*, putamen; *SurfArea*, cortical surface area; *Thal*, thalamus.

4. CONCLUSIONS

We used cross-trait structural equation modeling in a twin design to study the common genetic influences among three categories of common structural MRI and DTI brain measures: cortical thickness, subcortical volumes, and FA in WM tracts. We found significant genetic correlations between FA in several WM tracts and subcortical volume ROIs, notably the thalamus and pallidum. Association between cortical thickness and both volume ROIs and WM was not detectable, even in this relatively large sample.

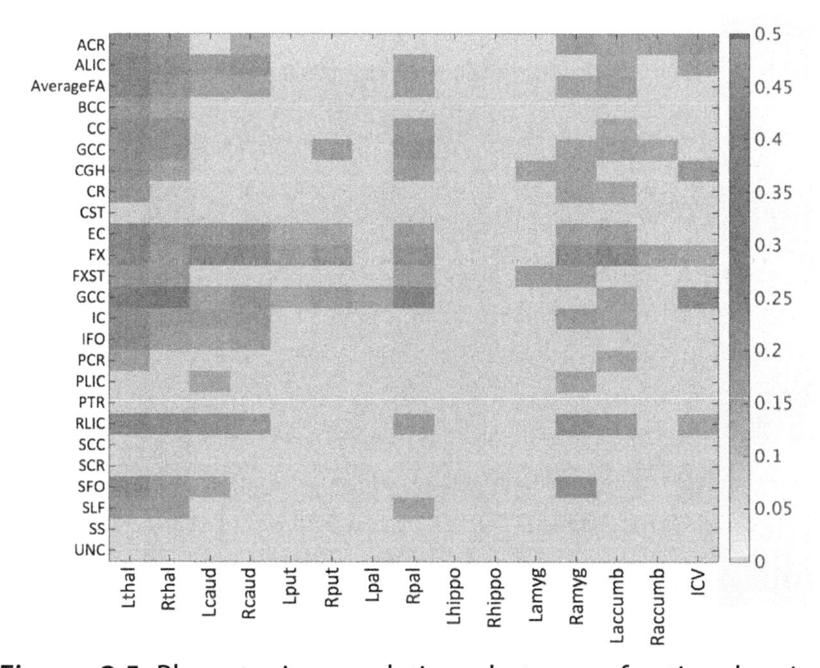

Figure 2.5 Phenotypic correlations between fractional anisotropy and volume measures. Colored elements (light gray in print versions) indicate significant associations after false discovery rate correction. Warmer colors (darker colors in print versions) indicate higher correlation (Pearson's *r*) values. *Accumb*, nucleus accumbens; *ACR*, anterior corona radiata; *ALIC*, anterior limb of the internal capsule; *Amyg*, amygdala; *BCC*, body of the corpus callosum; *Caud*, caudate; *CC*, corpus callosum; *CGC*, cingulum; *CHG*, hippocampal part of the cingulum; *CR*, corona radiata; *EC*, external capsule; *FXST*, fornix/stria terminalis; *GCC*, genu of the corpus callosum; *Hippo*, hippocampus; *IC*, internal capsule; *ICV*, intracranial volume; *Pal*, globus pallidus; *PCR*, posterior corona radiata; *PLIC*, posterior limb of the internal capsule; *PTR*, posterior thalamic radiation; *Put*, putamen; *RLIC*, retrolenticular part of the internal capsule; *SCC*, subcallosal cingulate white matter; *SCR*, superior region of corona radiata; *SFO*, superior fronto-occipital fasciculus; *SLF*, superior longitudinal fasciculus; *SS*, sagittal striatum; *SurfArea*, cortical surface area; *UNC*, uncinate fasciculus; *Thal*, thalamus.

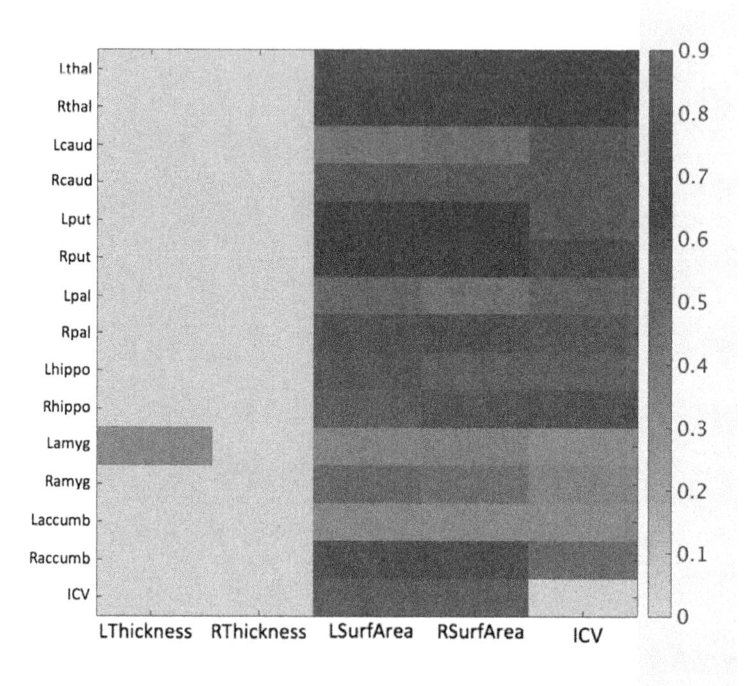

Figure 2.6 Phenotypic correlations between cortical thickness, surface area, and volume measures. Colored elements (light gray in print versions) indicate significant associations after FDR correction. Warmer colors (darker colors in print versions) indicate higher correlation (Pearson's *r*) values. Accumb, nucleus accumbens; *Amyg*, amygdala; *Caud*, caudate; *Hippo*, hippocampus; *ICV*, intracranial volume; *Pal*, globus pallidus; *Put*, putamen; *SurfArea*, cortical surface area; *Thal*, thalamus.

Even so, cortical surface area for each hemisphere was highly correlated with FA in several WM tracts and all of the subcortical volume measures.

Prior studies have suggested correlated genetic control across different brain structures [13,14]. Our finding of moderate genetic and phenotypic overlap between FA in several WM tracts and subcortical volumes is not surprising. Our strongest associations included the thalamus, pallidum, and several other basal ganglia structures. These structures are highly networked by WM tracts in cortico-basal ganglia-thalamocortical loops. Furthermore, they are physically closer to each other than to cortical gray matter.

Measures of thickness from the cortex—in addition to being structurally further away from subcortical regions—may also be affected by genetic programs that arose more recently in evolution. Cortical thickness is influenced largely by the number of neurons in a cortical column, according to the radial unit hypothesis of cortical development [15]. Surface area and thickness have different genetic origins [1], and MRI measurements of cortical thickness are not strongly associated with ICV or other head size measures [16]. This could partially explain the lack of association between cortical thickness and the other measures examined here.

ICV was genetically correlated with all subcortical volume measures, albeit modestly, with values ranging from 0.1 to 0.29. The phenotypic correlation between these measures was greater, ranging from 0.29 to 0.65. This may indicate that the relationship between these measures is not completely driven by common genetic factors.

There was some agreement between the genetic and phenotypic correlations, with the strongest phenotypic results being between FA in the left and right thalamus and the GCC. We did not detect an association between most of thickness measurements and any of the other phenotypes. In the future it may be useful to cluster regions that show genetic correlations (or even phenotypic correlations, as they tend to be quite similar).

Our primary aim was to identify genetically correlated—and conversely independent—regions across derived MRI measures that are commonly used in imaging genetics studies. These results can inform future gene discovery efforts, and clustering of these measures may boost power and allow dimensionality reduction. These correlation maps may also help elucidate the genetic control of the anatomy underlying common imaging measures.

Here we examined young adult subjects of European ancestry. Future studies of other populations would be valuable to test whether these genetic associations hold across different ethnicities.

GLOSSARY

White matter tracts analyzed

White matter tract	
ACR	Anterior *corona radiata*
ALIC	Anterior limb of the internal capsule
BCC	Body of the corpus callosum
CC	Corpus callosum
CGC	Cingulum
CHG	Hippocampal part of the cingulum
CR	*Corona radiata*
EC	External capsule
FXST	Fornix/*stria terminalis*
GCC	Genu of the corpus callosum
IC	Internal capsule
PCR	Posterior *corona radiata*
PLIC	Posterior limb of the internal capsule
PTR	Posterior thalamic radiation
RLIC	Retrolenticular part of the internal capsule
SCC	Subcallosal cingulate white matter
SCR	Superior region of *corona radiata*
SFO	Superior fronto-occipital fasciculus
SLF	Superior longitudinal fasciculus
SS	Sagittal striatum
UNC	Uncinate fasciculus

—cont'd

Regions where volume measurements were made	
Structure	
Thal	Thalamus
Caud	Caudate
Put	Putamen
Pal	Globus pallidus
Hippo	Hippocampus
Amyg	Amygdala
Accumb	Nucleus accumbens
ICV	Intracranial volume
SurfArea	Cortical surface area

REFERENCES

[1] A. Winkler, et al., Cortical thickness or grey matter volume? The importance of selecting the phenotype for imaging genetics studies, Neuroimage 53 (2009) 3.

[2] G. Blokland, G. Zubicaray, K. McMahon, M. Wright, Genetic and environmental influences on neuroimaging phenotypes: a meta-analytical perspective on twin imaging studies, Twin Research and Human Genetics 15 (2012).

[3] A. Fjell, H. Grydeland, S. Krogsrud, I. Amlien, D. Rohani, L. Ferschmann, A. Storsve, C. Tamnes, R. Sala-Llonch, P. Due-Tønnessen, A. Bjørnerud, A. Sølsnes, A. Håberg, J. Skranes, H. Bartsch, C.-H. Chen, W. Thompson, M. Panizzon, W. Kremen, A. Dale, K. Walhovd, Development and aging of cortical thickness correspond to genetic organization patterns, Proceedings of the National Academy of Sciences (2015), http://dx.doi.org/10.1073/pnas.201508831.

[4] P.M. Thompson, Cracking the brain's genetic code, Proceedings of the National Academy of Sciences (2015).

[5] M. Panizzon, C. Fennema-Notestine, L. Eyler, T. Jernigan, E. Prom-Wormley, M. Neale, K. Jacobson, M. Lyons, M. Grant, C. Franz, H. Xian, M. Tsuang, B. Fischl, L. Seidman, A. Dale, W. Kremen, Distinct genetic influences on cortical surface area and cortical thickness, Cerebral Cortex 19 (2009) 2728–2735.

[6] B.J. Casey, N. Tottenham, C. Liston, S. Durston, Imaging the developing brain: what have we learned about cognitive development? Trends in Cognitive Sciences 9 (2005) 104–110.
[7] D.P. Hibar, J.L. Stein, M.E. Renteria, A. Arias-Vasquez, S. Desrivières, N. Jahanshad, R. Toro, K. Wittfeld, L. Abramovic, M. Andersson, B.S. Aribisala, N.J. Armstrong, M. Bernard, M.M. Bohlken, M.P. Boks, J. Bralten, A.A. Brown, M.M. Chakravarty, Q. Chen, C.R. Ching, G. Cuellar-Partida, A. den Braber, S. Giddaluru, A.L. Goldman, O. Grimm, T. Guadalupe, J. Hass, G. Woldehawariat, A.J. Holmes, M. Hoogman, D. Janowitz, T. Jia, S. Kim, M. Klein, B. Kraemer, P.H. Lee, L.M. Olde Loohuis, M. Luciano, C. Macare, K.A. Mather, M. Mattheisen, Y. Milaneschi, K. Nho, M. Papmeyer, A. Ramasamy, S.L. Risacher, R. Roiz-Santiañez, E.J. Rose, A. Salami, P.G. Sämann, L. Schmaal, A.J. Schork, J. Shin, L.T. Strike, A. Teumer, M.M. van Donkelaar, K.R. van Eijk, R.K. Walters, L.T. Westlye, C.D. Whelan, A.M. Winkler, M.P. Zwiers, S. Alhusaini, L. Athanasiu, S. Ehrlich, M.M. Hakobjan, C.B. Hartberg, U.K. Haukvik, A.J. Heister, D. Hoehn, D. Kasperaviciute, D.C. Liewald, L.M. Lopez, R.R.R. Makkinje, M. Matarin, M.A. Naber, D.R. McKay, M. Needham, A.C. Nugent, B. Pütz, N.A. Royle, L. Shen, E. Sprooten, D. Trabzuni, S.S. van der Marel, K.J. van Hulzen, E. Walton, C. Wolf, L. Almasy, D. Ames, S. Arepalli, A.A. Assareh, M.E. Bastin, H. Brodaty, K.B. Bulayeva, M.A. Carless, S. Cichon, A. Corvin, J.E. Curran, M. Czisch, Common genetic variants influence human subcortical brain structures, Nature 520 (2015) 224–229.
[8] G. Zubicaray, M.-C. Chiang, K. McMahon, D. Shattuck, A. Toga, N. Martin, M. Wright, P. Thompson, Meeting the challenges of neuroimaging genetics, Brain Imaging and Behavior 2 (2008) 258–263.
[9] M. Neale, L. Cardon, Methodology for Genetic Studies of Twins and Families, 1992.
[10] N. Jahanshad, A. Lee, M. Barysheva, K. McMahon, G. Zubicaray, N. Martin, M. Wright, A. Toga, P. Thompson, Genetic influences on brain asymmetry: a DTI study of 374 twins and siblings, Neuroimage 52 (2010) 455–469.

[11] N. Jahanshad, O. Kohannim, D. Hibar, J. Stein, K. McMahon, G. Zubicaray, S. Medland, G. Montgomery, J. Whitfield, N. Martin, M. Wright, A. Toga, P. Thompson, Brain structure in healthy adults is related to serum transferrin and the H63D polymorphism in the HFE gene, Proceedings of the National Academy of Sciences 109 (2012) E851–E859.
[12] Y. Benjamini, Y. Hochberg, Controlling the false discovery rate: a practical and powerful approach to multiple testing, Journal of the Royal Statistical Society. Series B (Methodological) (1995) 289–300.
[13] D. Glahn, P. Thompson, J. Blangero, Neuroimaging endophenotypes: strategies for finding genes influencing brain structure and function, Human Brain Mapping 28 (2007) 488–501.
[14] J. Schmitt, L. Eyler, J. Giedd, W. Kremen, K. Kendler, M. Neale, Review of twin and family studies on neuroanatomic phenotypes and typical neurodevelopment, Twin Research and Human Genetics Official Journal of the International Society for Twin Study 10 (2007) 683–694.
[15] P. Rakic, A small step for the cell, a giant leap for mankind: a hypothesis of neocortical expansion during evolution, Trends in Neurosciences 18 (1995) 383–388.
[16] K. Im, J.-M. Lee, O. Lyttelton, S. Kim, A. Evans, S. Kim, Brain size and cortical structure in the adult human brain, Cerebral Cortex (New York, N.Y.: 1991) 18 (2008) 2181–2191.

CHAPTER THREE

Integration of Network-Based Biological Knowledge With White Matter Features in Preterm Infants Using the Graph-Guided Group Lasso

Michelle L. Krishnan[1], Zi Wang[2], Matt Silver[3], James P. Boardman[4], Gareth Ball[1], Serena J. Counsell[1], Andrew J. Walley[2], David Edwards[1], Giovanni Montana[1]

[1]King's College London, London, United Kingdom
[2]Imperial College London, London, United Kingdom
[3]London School of Hygiene and Tropical Medicine, London, United Kingdom
[4]University of Edinburgh, Edinburgh, United Kingdom

Contents

Abstract

The effect of prematurity on normal developmental programs of white and gray matter as evaluated with magnetic resonance imaging indicates global changes in white and gray matter with functional implications. We have previously identified an association between lipids and diffusion tensor imaging features in

Imaging Genetics
ISBN: 978-0-12-813968-4
http://dx.doi.org/10.1016/B978-0-12-813968-4.00003-1

preterm infants, both through a candidate gene approach and a data-driven statistical genetics method. Here we apply a penalized linear regression model, the graph-guided group lasso (GGGL), that can utilize prior knowledge and select single nucleotide polymorphisms (SNPs) within functionally related genes associated with the trait. GGGL incorporates prior information from SNP-gene mapping as well as from the gene functional interaction network to guide variable selection.

Keywords: Brain development; Machine learning; Magnetic resonance imaging; PPARG; Sparse regression

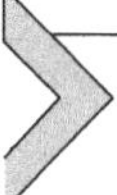

1. BACKGROUND AND AIMS

The incidence of preterm birth is increasing steadily [1], with many survivors experiencing adverse motor, cognitive, and psychiatric sequelae [2]. Over 30% have cognitive problems; 30% attention deficit disorder; and 5% develop autistic spectrum disorders, posing a "growing and neglected problem" for public health [3,4]. The effect of prematurity on normal developmental programs of white and gray matter as evaluated with magnetic resonance imaging (MRI) indicates global changes in white matter with functional implications [5–7], accompanied by gray matter alterations [8–11].

Diffusion tensor imaging (DTI) provides measures of white matter microstructure that are correlated with neurodevelopmental outcome [12] and highly heritable [13]. Decreased fractional anisotropy (FA) in preterm infants is related to cognitive, fine-motor and gross-motor outcome at 2 years [6], with cognitive consequences extending into adulthood in very preterm individuals [14].

We have previously identified an association between lipids and DTI features in preterm infants, both through a candidate gene approach [15] and a data-driven statistical genetics method [16,17]. This highlighted the KEGG peroxisome proliferator-activated receptor pathway (PPAR) as most highly ranked in association with the white matter phenotype. Here we apply a penalized linear regression model, the graph-guided group lasso (GGGL) [18], that can utilize such prior knowledge and select single nucleotide polymorphisms (SNPs) within functionally related genes associated with the trait. GGGL incorporates prior information from SNP-gene mapping as well as from the gene functional interaction network to guide variable selection: SNPs are grouped into genes, and genes are organized into a weighted gene network encoding the functional relatedness between all pairs of genes. The model selects functionally related genes and SNPs within these genes that are associated with a quantitative trait, using graph and grouping structure on hierarchical biological variants to drive variable selection at multiple levels.

The rest of this paper is organized as follows: In Section 2 we briefly recap the GGGL model as detailed in Ref. [18]. Analysis procedures are given in Section 3, and the results are presented in Section 4. We conclude in Section 5.

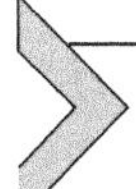

2. GRAPH-GUIDED GROUP LASSO

Let X denote the n-by-p design matrix containing n independent subjects for which p SNPs have been recorded, and let y denote an n-dimensional real vector consisting of the univariate quantitative trait. We assume a linear predictive model of X on y. By normalizing the columns of X to have zero sum and unit length and centering y, we can drop the intercept term so that $y = X\beta + \varepsilon$, where β consists of the linear coefficients, and ε is the random error.

In addition to the design matrix, from the prior information SNPs are arranged into mutually exclusive groups $R = R_1, R_2, R_3\ldots$ by mapping to their nearest genes. Let $G = G(V,E)$ be the given network where the vertex set V corresponds to the set of genes, and the edge set E corresponds to the pairs of genes that are known to share biological functions. Suppose there are $|R|$ genes involved in the study, we assume the pairwise relatedness between the genes is represented by the $|R|$-by-$|R|$ matrix W, where W_{IJ} can be either binary or a real number between 0 and 1 such that a larger weight means gene R_I and R_J are more likely to be involved in the same biological process (BP).

The GGGL linear coefficients β are estimated by minimizing the standard least square loss plus a composite penalty term as below:

$$\frac{1}{2}\|y - X\beta\|_2^2 + \lambda_1 \sum_{R_K \in R} \sqrt{|R_K|}\|\beta_K\|_2 + \lambda_2\|\beta\|_1$$

$$+\frac{1}{2}\mu \sum_{i\in R_I, j\in R_J, I\sim J} W_{IJ}\left(\beta_i - \beta_j\right)^2 \tag{3.1}$$

where $|R_K|$ denotes the size of group R_K, β_K denotes the coefficients corresponding to the SNPs in R_K, and $I \sim J$ denotes the pair of genes R_I and R_J for which W_{IJ} is nonzero. λ_1, λ_2, and μ are nonnegative regularization parameters to balance the weight each term is accorded. Notably when $\lambda_1 = \lambda_2 = \mu = 0$, (Eq. 3.1) reduces to a standard linear regression; when only $\lambda_1 = \mu = 0$, (Eq. 3.1) reduces to lasso regression, which uniformly shrinks the coefficients toward zero and sets some of them to exactly zero, thus attaining sparse model in which only the most important SNPs are selected to predict the response [19]; when only $\mu = 0$, (Eq. 3.1) reduces to the sparse group lasso regression, which selects most predictive groups (genes) and identifies the important SNPs within the selected genes [20].

In GGGL, sparsity is imposed by the second and third term in (Eq. 3.1), where the number of genes and SNPs selected is regularized by λ_1 and λ_2, respectively: for fixed λ_2, as λ_1 increases from zero, less genes are selected; likewise, controlling λ_1 so that a fixed number of genes are selected, increasing λ_2 will result in less SNPs being selected within the set of selected genes. The last term in (Eq. 3.1) is a Laplacian penalty penalizing the squared difference between all functionally related SNPs. This penalty smoothens the coefficients β_i and β_j toward the same value between their original estimates obtained without the penalty [21], thus encouraging the functionally related SNPs to be selected into or left out of the model altogether. The regularization parameter μ leverages the contribution from prediction errors, and the prior knowledge encoded by the network. As μ tends to infinity, all functionally related SNPs obtain the same coefficients, which gives rise to the lowest prediction error. From another prospect, μ can be regarded as a regularizer that can regulate the distribution of selected genes/SNPs in the given network. Suppose we tune λ_1 and λ_2 such that the same number of genes/SNPs is selected for various μ's. When $\mu = 0$, the selected groups/features may be spread all over the place in the network, depending on the signal-to-noise ratio and how well the network captures the underlying BP, which leads to the observed variability in the quantitative trait. However, as μ moves away from zero, the selected genes/SNPs become more clustered in the network, and they tend to lie in one or a few connected subnetworks ("components").

For computation we have worked out a parallelized estimation algorithm of the GGGL, based on Ref. [22] and as detailed in Ref. [18]. At each iteration, the algorithm randomly takes a subset of groups and updates the coefficients in parallel using a coordinate descent method. We implemented this algorithm on graphical processing units to so as to benefit from the thousands of Compute Unified Device Architecture cores available when dealing with high-dimensional data.

3. ANALYSIS

3-T magnetic resonance images and saliva were acquired for 72 preterm infants (mean gestational age 28 + 4 weeks, mean postmenstrual age at scan 40 + 3 weeks). Imaging was performed on a Philips 3-T system (Philips Medical Systems, Netherlands) using an eight-channel phased-array head coil. Single-shot echo-planar diffusion tensor imaging was acquired in the transverse plane in 15 noncollinear directions.

Salivary DNA samples were genotyped on Illumina HumanOmniExpress-12 arrays. The genotype matrix was recoded in terms of minor allele counts, including only SNPs with minor allele frequency ≥5% and ≥99% genotyping rate [23]. The functional relationships between genes within the (PPAR) signaling pathway were systematically described by clustering the genes based on their Gene Ontology (GO) BP annotations [24]. This resulted in an adjacency matrix based on pairwise semantic similarity of GO

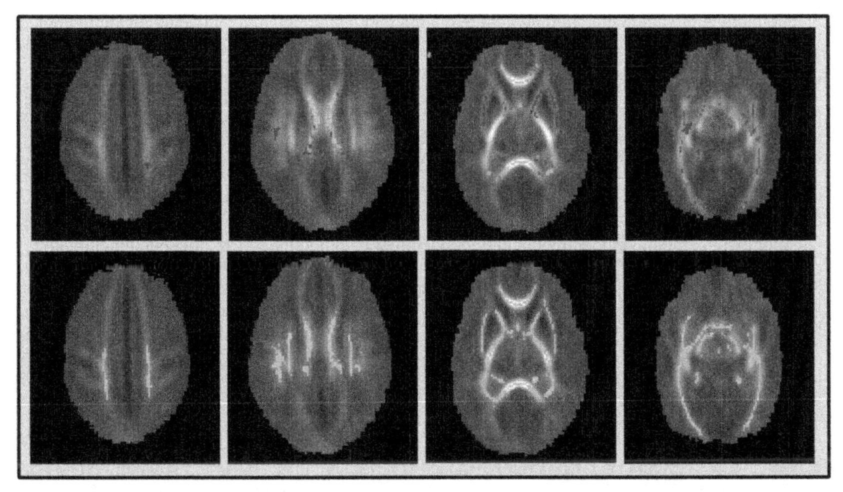

Figure 3.1 Group white matter diffusion tensor imaging skeleton, showing voxels that vary significantly between individuals (corrected $P < .05$ for all voxels, darker blue (gray in print versions): lower P-value). Axial views superior to inferior left to right. Top row: voxels varying between individuals adjusting for postmenstrual age (PMA) at scan. Bottom row: voxels varying between individuals adjusting for gestational age at birth and PMA.

terms, which has been shown to correlate with gene expression [25], protein sequence similarity [26], and protein family similarity [27].

FA maps were constructed from 15-direction DTI, and Tract-based spatial statistics [28] was used to obtain a group white matter skeleton varying with degree of prematurity, adjusting for age at scan (Fig. 3.1). The phenotype was reduced to its first principal component, and GGGL was applied to the genes in the PPAR pathway (Fig. 3.2).

The regularization parameters λ_1, λ_2, and μ in GGGL are traditionally tuned by a cross-validation procedure in which the optimal parameters are chosen to minimize out-of-sample prediction errors. However, parameters tuned using this criterion do not necessarily give rise to a

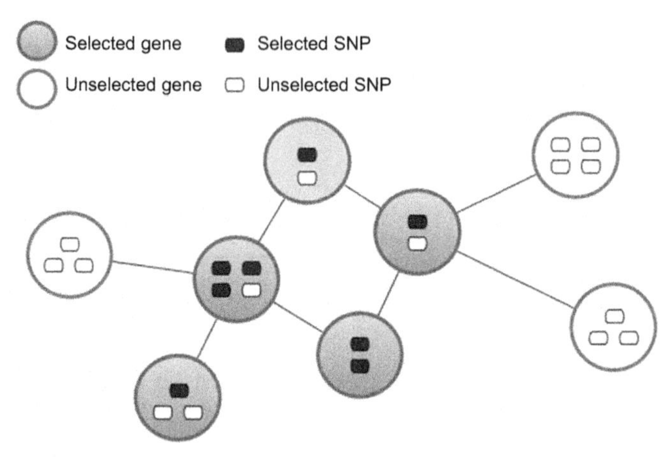

Figure 3.2 Sparsity patterns obtained by the graph-guided group lasso model: Single nucleotide polymorphisms (SNPs) are grouped into genes, and the pairwise functional relatedness between genes is represented by a network. The model selects genes associated with the univariate quantitative trait and the most influential SNPs within these genes. The prior structural knowledge guides the model to select genes which are connected in the network.

model comprising the genes and SNPs that are truly associated with the quantitative trait, which is the main objective of our analysis. In fact, it has been observed that the parameters giving the best out-of-sample prediction performance often result in a larger model than the true one, where many noise variables are recruited [29]. Here we adopted a data subsampling approach called stability selection [30], which directly addresses the problem of variable ranking and variable selection according to their associations with the response. Specifically, for a fixed value of μ, we randomly extracted a half of the samples, without replacement, and fitted GGGL on the extracted data. We repeated this procedure 1000 times, and for each subsampled dataset we tuned λ_1 and λ_2 in GGGL so that we always selected about 20% of the genes and half of the SNPs within selected genes. We then computed the empirical selection probability of each SNP using the variable selection results obtained in all 1000 subsampled datasets. As μ controls the weight of the prior knowledge, which may lead to very different distribution patterns of the selected genes/SNPs in the network, we repeated the stability selection procedure for three different μ's: 0.1, 1, and 10. These different settings resulted in the same set of genes with selection probability >0.4 (a threshold representing a step change in the probability distribution), indicating robustness of the results. Genes with a selection probability threshold >0.4 are listed in Table 3.1.

4. RESULTS

Within the PPAR pathway, the GGGL method selected a subset of highly ranked genes (5/69) functionally related in terms of GO BP and linearly correlated with white matter FA. These were aquaporin 7 (AQP7), malic enzyme 1, NADP(+)-dependent, cytosolic (ME1), perilipin 1 (PLIN1), solute carrier family 27 (fatty acid transporter), member 1 (SLC27A1), and acetyl-CoA acyltransferase 1 (ACAA1). Analysis of transcriptional regulation using the PASTAA algorithm [31] indicated that ACAA1, AQP7, ME1, and SLC27A1 are jointly regulated by the EGR-4 transcription factor (adjusted *P*-value 7.7×10^{-4}).

All five of these genes are involved in BPs that are relevant to brain development in preterm infants and the neurometabolic sequelae they experience in later life. ME1 appears to be involved in gene regulation in infants as a result of maternal obesity [32], and PLIN1 regulates the

Table 3.1 Top ranked single nucleotide polymorphisms mapped to genes, sorted by selection probability

SNP$Names	Gene$mapping	μ = 10	μ = 1.0	μ = 0.1
rs3758267	AQP7	0.855	0.841	0.86
rs4879696	AQP7	0.817	0.804	0.826
rs1143796	ME1	0.684	0.741	0.688
rs1535588	ME1	0.684	0.741	0.688
rs1170348	ME1	0.683	0.741	0.688
rs1144184	ME1	0.682	0.74	0.683
rs9449593	ME1	0.659	0.717	0.663
rs3798890	ME1	0.648	0.698	0.644
rs6917851	ME1	0.634	0.679	0.62
rs12191369	ME1	0.634	0.679	0.62
rs1180242	ME1	0.607	0.639	0.609
rs1180192	ME1	0.593	0.626	0.605
rs7169981	PLIN1	0.553	0.566	0.571
rs11073884	PLIN1	0.553	0.566	0.571
rs2289487	PLIN1	0.552	0.565	0.572
rs8179043	PLIN1	0.546	0.558	0.568
rs12351969	AQP7	0.537	0.526	0.552
rs6790738	ACAA1	0.533	0.582	0.547
rs12630114	ACAA1	0.532	0.581	0.549
rs7744	ACAA1	0.527	0.574	0.542
rs6512198	SLC27A1	0.515	0.559	0.536
rs11086076	SLC27A1	0.515	0.565	0.536
rs9825655	ACAA1	0.513	0.561	0.533
rs2278280	SLC27A1	0.511	0.56	0.535
rs11665931	SLC27A1	0.506	0.54	0.527
rs4808657	SLC27A1	0.49	0.539	0.517
rs11668681	SLC27A1	0.483	0.531	0.504
rs7255307	SLC27A1	0.465	0.503	0.481
rs9311180	ACAA1	0.459	0.489	0.471
rs1954537	ME1	0.458	0.519	0.47
rs4808652	SLC27A1	0.455	0.483	0.464
rs6599263	ACAA1	0.448	0.485	0.468

Table 3.1 Top ranked single nucleotide polymorphisms mapped to genes, sorted by selection probability—cont'd

SNP$Names	Gene$mapping	$\mu = 10$	$\mu = 1.0$	$\mu = 0.1$
rs1123569	ACAA1	0.438	0.474	0.459
rs13219666	ME1	0.422	0.442	0.434
rs11670276	SLC27A1	0.415	0.453	0.426
rs10890467	CYP4A22	0.409	0.43	0.417
rs750385	ME1	0.403	0.43	0.425
rs11666579	SLC27A1	0.403	0.425	0.431

ACAA1, acetyl-CoA acyltransferase 1; *AQP7*, aquaporin 7; *PLIN1*, perilipin 1.

function of immune cells in the brain (microglia) [33]. SLC27A1 (also known as fatty acid transport molecule 1, FATP1) is involved in fatty acid transport across the blood brain barrier [34], AQP7 expression is associated with insulin resistance and obesity [35], and ACAA1 is involved in neuronal growth and myelinogenesis [36].

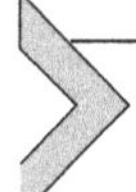

5. CONCLUSIONS

GGGL analysis of genes in the PPAR lipid pathway identified five highly ranked genes involved in neuronal growth, myelinogenesis, and nervous system response to inflammation [33,34,36], which are jointly transcriptionally regulated by a transcription factor that is also important in neurodevelopment [37]. Given that preterm infants are known to be at increased risk of both mental illness and cardiovascular morbidity in later life [38], this work leverages a data-driven strategy to propose a unifying mechanism through which these systemic effects might be mediated and suggests a promising avenue for therapeutic intervention.

REFERENCES

[1] H. Blencowe, S. Cousens, M.Z. Oestergaard, D. Chou, A.B. Moller, R. Narwal, A. Adler, C. Vera Garcia, S. Rohde, L. Say, J.E. Lawn, National, regional, and worldwide estimates of preterm birth rates in the year 2010 with time trends since 1990 for selected countries: a systematic analysis and implications, Lancet 379 (2012) 2162–2172.
[2] S. Johnson, N. Marlow, Preterm birth and childhood psychiatric disorders, Pediatric Research 69 (2011) 11R–18R.
[3] S. Saigal, L.W. Doyle, An overview of mortality and sequelae of preterm birth from infancy to adulthood, Lancet 371 (2008) 261–269.
[4] T. Moore, E.M. Hennessy, J. Myles, S.J. Johnson, E.S. Draper, K.L. Costeloe, N. Marlow, Neurological and developmental outcome in extremely preterm children born in England in 1995 and 2006: the EPICure studies, BMJ: British Medical Journal 345 (2012) e7961.
[5] L. Eikenes, G.C. Lohaugen, A.M. Brubakk, J. Skranes, A.K. Haberg, Young adults born preterm with very low birth weight demonstrate widespread white matter alterations on brain DTI, Neuroimage 54 (2011) 1774–1785.
[6] B.J. van Kooij, L.S. de Vries, G. Ball, I.C. van Haastert, M.J. Benders, F. Groenendaal, S.J. Counsell, Neonatal tract-based spatial statistics findings and outcome in preterm infants, AJNR: American Journal of Neuroradiology 33 (2012) 188–194.
[7] M. Groppo, D. Ricci, L. Bassi, N. Merchant, V. Doria, T. Arichi, J.M. Allsop, L. Ramenghi, M.J. Fox, F.M. Cowan, S.J. Counsell, A.D. Edwards, Development of the optic radiations and visual function after premature birth, Cortex: A Journal Devoted to the Study of the Nervous System and Behavior 56 (2014) 30–37.
[8] G. Ball, L. Srinivasan, P. Aljabar, S.J. Counsell, G. Durighel, J.V. Hajnal, M.A. Rutherford, A.D. Edwards, Development of cortical microstructure in the preterm human brain, Proceedings of the National Academy of Sciences of the United States of America 110 (2013) 9541–9546.
[9] J. Vinall, R.E. Grunau, R. Brant, V. Chau, K.J. Poskitt, A.R. Synnes, S.P. Miller, Slower postnatal growth is associated with delayed cerebral cortical maturation in preterm newborns, Science Translational Medicine 5 (2013) 168ra168.

[10] T.A. Smyser, C.D. Smyser, C.E. Rogers, S.K. Gillespie, T.E. Inder, J.J. Neil, Cortical gray and adjacent white matter demonstrate synchronous maturation in very preterm infants, Cerebral Cortex (2015).
[11] M. Ajayi-Obe, N. Saeed, F.M. Cowan, M.A. Rutherford, A.D. Edwards, Reduced development of cerebral cortex in extremely preterm infants, Lancet 356 (2000) 1162–1163.
[12] S.J. Counsell, A.D. Edwards, A.T. Chew, M. Anjari, L.E. Dyet, L. Srinivasan, J.P. Boardman, J.M. Allsop, J.V. Hajnal, M.A. Rutherford, F.M. Cowan, Specific relations between neurodevelopmental abilities and white matter microstructure in children born preterm, Brain: A Journal of Neurology 131 (2008) 3201–3208.
[13] X. Geng, E.C. Prom-Wormley, J. Perez, T. Kubarych, M. Styner, W. Lin, M.C. Neale, J.H. Gilmore, White matter heritability using diffusion tensor imaging in neonatal brains, Twin Research and Human Genetics: The Official Journal of the International Society for Twin Studies 15 (2012) 336–350.
[14] M.P. Allin, D. Kontis, M. Walshe, J. Wyatt, G.J. Barker, R.A. Kanaan, P. McGuire, L. Rifkin, R.M. Murray, C. Nosarti, White matter and cognition in adults who were born preterm, PLoS One 6 (2011) e24525.
[15] J.P. Boardman, A. Walley, G. Ball, P. Takousis, M.L. Krishnan, L. Hughes-Carre, P. Aljabar, A. Serag, C. King, N. Merchant, L. Srinivasan, P. Froguel, J. Hajnal, D. Rueckert, S. Counsell, A.D. Edwards, Common genetic variants and risk of brain injury after preterm birth, Pediatrics 133 (2014) e1655–1663.
[16] M. Silver, G. Montana, Alzheimer's Disease Neuroimaging Initiative, Fast identification of biological pathways associated with a quantitative trait using group lasso with overlaps, Statistical Applications in Genetics and Molecular Biology 11 (2012). Article 7.
[17] M.L. Krishnan, J.P. Boardman, M. Silver, G. Ball, S.J. Counsell, A.J. Walley, A.D. Edwards, G. Montana, Investigation of biological pathways involved in brain development in preterm neonates using a multivariate phenotype and sparse regression, in: MICCAI Workshop on Imaging Genetics, 2014.
[18] Z. Wang, G. Montana, The graph-guided group lasso for genome-wide association studies, in: J.A.K. Suykens, M. Signoretto, A. Argyrou (Eds.), Regularization, Optimization, Kernels, and Support Vector Machines, CRC Press, 2014.

[19] R. Tibshirani, Regression shrinkage and selection via the lasso, Journal of the Royal Statistical Society, Series B 58 (1996) 267–288.
[20] N. Simon, J. Friedman, T. Hastie, R. Tibshirani, A sparse-group lasso, Journal of Computational and Graphical Statistics 22 (2013) 231–245.
[21] C. Li, H. Li, Network-constrained regularization and variable selection for analysis of genomic data, Bioinformatics 24 (2008) 1175–1182.
[22] P. Richtárik, M. Takáč, Parallel coordinate descent methods for big data optimization, ArXiv 1212.0873v1212, 2013.
[23] S. Purcell, B. Neale, K. Todd-Brown, L. Thomas, M.A. Ferreira, D. Bender, J. Maller, P. Sklar, P.I. de Bakker, M.J. Daly, P.C. Sham, PLINK: a tool set for whole-genome association and population-based linkage analyses, American Journal of Human Genetics 81 (2007) 559–575.
[24] K. Ovaska, M. Laakso, S. Hautaniemi, Fast gene ontology based clustering for microarray experiments, BioData Mining 1 (2008) 11.
[25] P. Khatri, S. Draghici, Ontological analysis of gene expression data: current tools, limitations, and open problems, Bioinformatics 21 (2005) 3587–3595.
[26] R. Apweiler, A. Bairoch, C.H. Wu, W.C. Barker, B. Boeckmann, S. Ferro, E. Gasteiger, H. Huang, R. Lopez, M. Magrane, M.J. Martin, D.A. Natale, C. O'Donovan, N. Redaschi, L.S. Yeh, UniProt: the universal protein knowledgebase, Nucleic Acids Research 32 (2004) D115–D119.
[27] F.M. Couto, M.J. Silva, P.M. Coutinho, Measuring semantic similarity between gene ontology terms, Data and Knowledge Engineering 61 (2006) 137–152.
[28] S.M. Smith, M. Jenkinson, H. Johansen-Berg, D. Rueckert, T.E. Nichols, C.E. Mackay, K.E. Watkins, O. Ciccarelli, M.Z. Cader, P.M. Matthews, T.E. Behrens, Tract-based spatial statistics: voxelwise analysis of multi-subject diffusion data, Neuroimage 31 (2006) 1487–1505.
[29] C. Leng, Y. Lin, G. Wahba, A note on the lasso and related procedures in model selection, Statistica Sinica 16 (2006) 1273–1284.

[30] M. Meinshausen, P. Bühlmann, Stability selection, Journal of the Royal Statistical Society, Series B 72 (2010) 417–473.
[31] H.G. Roider, T. Manke, S. O'Keeffe, M. Vingron, S.A. Haas, PASTAA: identifying transcription factors associated with sets of co-regulated genes, Bioinformatics 25 (2009) 435–442.
[32] M. Dahlhoff, S. Pfister, A. Blutke, J. Rozman, M. Klingenspor, M.J. Deutsch, B. Rathkolb, B. Fink, M. Gimpfl, M. Hrabe de Angelis, A.A. Roscher, E. Wolf, R. Ensenauer, Periconceptional obesogenic exposure induces sex-specific programming of disease susceptibilities in adult mouse offspring, Biochimica et biophysica acta 1842 (2014) 304–317.
[33] A. Khatchadourian, S.D. Bourque, V.R. Richard, V.I. Titorenko, D. Maysinger, Dynamics and regulation of lipid droplet formation in lipopolysaccharide (LPS)-stimulated microglia, Biochimica et biophysica acta 1821 (2012) 607–617.
[34] R.W. Mitchell, N.H. On, M.R. Del Bigio, D.W. Miller, G.M. Hatch, Fatty acid transport protein expression in human brain and potential role in fatty acid transport across human brain microvessel endothelial cells, Journal of Neurochemistry 117 (2011) 735–746.
[35] J. Lebeck, Metabolic impact of the glycerol channels AQP7 and AQP9 in adipose tissue and liver, Journal of Molecular Endocrinology 52 (2014) R165–R178.
[36] S. Houdou, S. Takashima, Y. Suzuki, Immunohistochemical expression of peroxisomal enzymes in developing human brain, Molecular and Chemical Neuropathology 19 (1993) 235–248.
[37] J.J. Volpe, Brain injury in premature infants: a complex amalgam of destructive and developmental disturbances, Lancet Neurology 8 (2009) 110–124.
[38] E. Bayman, A.J. Drake, C. Piyasena, Prematurity and programming of cardiovascular disease risk: a future challenge for public health? Archives of disease in childhood, Fetal and Neonatal Edition 99 (2014) F510–F514.

Classifying Schizophrenia Subjects by Fusing Networks From Single-Nucleotide Polymorphisms, DNA Methylation, and Functional Magnetic Resonance Imaging Data

Su-Ping Deng[1,2] De-Shuang Huang[2], Dongdong Lin[3], Vince D. Calhoun[3,4], Yu-Ping Wang[1]

[1]Tulane University, New Orleans, LA, United States
[2]Tongji University, Shanghai, China
[3]Mind Research Network, Albuquerque, NM, United States
[4]The University of New Mexico, Albuquerque, NM, United States

Contents

Imaging Genetics
ISBN: 978-0-12-813968-4
http://dx.doi.org/10.1016/B978-0-12-813968-4.00004-3

Abstract

To comprehensively employ complementary information from multiple types of data for better disease diagnosis, in this study, we applied a network fusion–based approach to three types of data including genetic, epigenetic, and neuroimaging data from a study of schizophrenia patients (SZ). A network is a map of interactions, which is helpful for investigating the connectivity of components or links between subunits. We exploit the potential of using networks as features for discriminating SZ from healthy controls. We first constructed a single network for each type of data. Then we built four fused networks by network fusion method: three fused networks for each combination of two types of data and one fused network for all of three data sets. Based on the local consistency of network, we can predict the group of SZ subjects with unknown labels. The group prediction method was applied to test the power of network-based features and the performance was evaluated by a tenfold cross-validation. The results show that the prediction accuracy is the highest when applying our prediction method to the fused network derived from three data types among all seven tested networks. As a conclusion, when making a diagnosis or predicting the labels of SZ subjects, we recommend more approaches that attempt to comprehensively utilize the multiple types data that are often available.

Keywords: Classification; Network fusion; Schizophrenia; Spectral clustering

1. INTRODUCTION

Schizophrenia (SZ) is a chronic, severe mental disorder showing a variety of symptoms such as false beliefs, unclear or confused thinking, and auditory hallucinations. SZ patients may not make sense when they talk, and some of them even may sit for hours without moving or talking. SZ is usually diagnosed based on criteria set in either the American Psychiatric Association's fifth edition of the Diagnostic and Statistical Manual of Mental Disorders (DSM-5), or the World Health Organization's International Statistical Classification of Diseases and Related Health Problems (ICD-10). These criteria use the self-reported experiences of the patient and reported abnormalities in behaviors, followed by a clinical assessment by a mental health professional. Symptoms associated with SZ occur along a continuum in the population and must reach a certain severity before a diagnosis is made [1]. So far there is no objective test [2]. Although in psychiatry as in all of general medicine there is an irreducible element of the subjective [3], we should try our best to reduce the subjective part of medical and psychiatry practice. This goal is well recognized by the National Institutes of Mental Health as well [4].

In this study, we explore the potential of using connectivity networks of patients to predict the diagnosis. Recently there has been much research on networks in the biomedical and bioinformatics field [5—8]. However, there is little research about network construction and analysis from multiple types of biology data. A network demonstrates a system consists of a collection of subunits (nodes) as well as their links, such as species units linked into a whole food web. In a topological sense, a network is a set of nodes and a set of directed or undirected edges between the nodes.

Networks focus on the organization of the system rather than on its components. So we can exploit the features of networks to classify SZ subjects.

Rapid advances of high-throughput technology are making a flood of individual data sets, including genomic, epigenomic, transcriptomic, and proteomic information. Each type of data offers a unique and complementary insight on genome organization and cellular function. Therefore, researchers can take advantage of diverse types of data and combine them to create an integrative view of a disease or a biological process. Beyer et al. [10] suggested that both genetic and physical data are essential to our understanding of biological systems. Genetic data explain the "what" of biological systems: what is the function of a gene, what is its phenotype, and what is its target? Physical data explain the "how": how does a gene or protein execute its function? There are several challenges in integrating physical and genetic networks, [9,10]. In the past decades, there has been a lot of work to integrate multitypes of data to answer biological questions about complex diseases. Zhang et al. [11] constructed a causal network with an integrative network-based approach, which presents a framework to test models of disease mechanisms underlying late-onset Alzheimer's disease. Lin and Cao et al. have conducted research on combining imaging and single-nucleotide polymorphisms (SNPs) data based on sparse regression model [12–16] to explore the correspondence between the diverse features and also combine them to improve the disease classification.

In our current study, we integrated three types of data: SNPs, DNA methylation, and functional MRI (fMRI) for the construction of networks, which will then be fused to classify SZ subjects. We first constructed three single networks separately and then fused two or three single networks using a network fusion method. As a result, we created four fused networks: three fused networks from the combinations of two data modalities and one fused network for three data modalities. Based on the fused networks, we clustered the SZ patients into several groups and predicted the group to which new subjects belong, i.e., identifying whether a subject is a SZ patient.

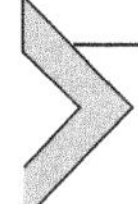

2. MATERIALS AND METHODS

2.1 Data Sets

In this study, participant recruitment and data collection were conducted at the Mind Research Network. Three types of data (SNPs, DNA methylation, and fMRI) were collected from 208 subjects including 96 SZ patients (age: 34 ± 11, 22 females) and 112 healthy controls (age: 32 ± 11, 44 females). All of them were provided written informed consents. Healthy participants were free of any medical, neurological, or psychiatric illnesses and had no history of substance abuse. By the clinical interview of patients for DSM-IV-TR disorders or the Comprehensive Assessment of Symptoms and History, patients met criteria for DSM-IV-TR SZ. Antipsychotic history was collected as part of the psychiatric assessment. After a series of quality controls (QC), we selected 184 subjects, including 80 SZ cases (age: 34 ± 11, 20 females and 60 males) and 104 healthy controls (age: 32 ± 11, 38 females and 66 males). After preprocessing, 27,508 DNA methylation sites, 41,236 fMRI voxels, and 722,177 SNPs loci were obtained for the subsequent biomarker selections [15,17–21].

2.1.1 Single-Nucleotide Polymorphisms Data Collection and Preprocessing

A blood sample was obtained for each participant, and DNA was extracted. Genotyping for all participants was performed at the Mind Research Network using the Illumina Infinium HumanOmni1-Quad assay covering 1,140,419 SNP loci. Bead Studio was used to make the final genotype calls. Next, the PLINK software package (http://pngu.mgh.harvard.edu/~purcell/plink) was used to perform a series of standard quality control procedures, resulting in the final data set spanning 722,177 SNP loci. Each SNP was categorized into three clusters based on their genotype and was represented with discrete numbers: 0 for "BB" (no minor allele), 1 for "AB" (one minor allele), and 2 for "AA" (two minor alleles).

2.1.2 Functional Magnetic Resonance Imaging Data Collection and Preprocessing

The fMRI data were collected during a sensorimotor task; a block-design motor response to auditory stimulation. During the on-block, 200 ms tones were presented with a 500 ms stimulus-onset asynchrony. A total of 16 different tones were presented in each on block, with frequency ranging from 236 to 1318 Hz. The fMRI images were acquired on Siemens 3T Trio Scanners and a 1.5 T Sonata with echo-planar imaging sequences using the following parameters (TR = 2000 ms, TE = 30 ms (3.0 T)/40 ms (1.5 T), field of view = 22 cm, slice thickness = 4 mm, 1 mm skip, 27 slices, acquisition matrix = 64×64, flip angle = 90 degree). Data were preprocessed in SPM5 (http://www.fil.ion.ucl.ac.uk/spm) and were realigned, spatially normalized, and resliced to $3 \times 3 \times 3$ mm, smoothed with a $10 \times 10 \times 10$ mm^3 Gaussian kernel to reduce spatial noise, and analyzed by multiple regression considering the stimulus and their temporal derivatives plus an intercept term as regressors. Finally the stimulus-on versus stimulus-off contrast images were extracted with $53 \times 63 \times 46$ voxels, and all the voxels with missing measurements were excluded.

2.1.3 DNA Methylation Data Collection and Preprocessing

DNA from blood samples was assessed by the Illumina Infinium Methylation27 assay. A methylation value, beta (β), represents the ratio of the methylated probe intensity to the total probe intensity. A series of QC on the beta values were applied to remove bad samples and probes, such as (1) Beta value QC: Change any beta value to not a number (NaN), if $P > .05$. (2) Bad sample/bad marker removing: Samples with >5% of missing (NaN) values; markers with >5% of missing (NaN) values. This resulted in the identification of good methylation data from 224 subjects, 27,508 markers (some have missing values <5%). After QC, we used the K nearest neighbor (KNN) method to impute for the missing values.

2.2 Methods

2.2.1 Similarity Network Fusion

Here, we employed the similarity network fusion (SNF) method proposed by Wang et al. [22], for which an R package SNFtool [22] is also available for use. The SNF is inspired by the theoretical multiview learning framework, which was developed for the computer vision and image processing applications [23]. The SNF constructs fused networks of samples by comparing samples' molecular (or phenotypic) profiles. The fused networks are then used for subtyping and label prediction.

Suppose we have n samples (e.g., patients) and m measurements (e.g., DNA methylation). A patient similarity network is represented as a graph $G=(V,E)$. The vertices V correspond to the patients $\{x_1, x_2, \ldots, x_n\}$, and the edges E are the weighted value of the similarity between patients. Edge weights are represented by an $n \times n$ similarity matrix $\mathbf{W}$, with each $\mathbf{W}(i,j)$ indicating the similarity between patients x_i and x_j. $\rho(x_i,x_j)$ is represented as the Euclidean distance between patients x_i and x_j. A scaled exponential kernel is used to determine the weight of the edge:

$$\mathbf{W}(i,j) = \exp\left(-\frac{\rho^2(x_i, x_j)}{\mu\,\varepsilon_{i,j}}\right) \tag{4.1}$$

where μ is a hyperparameter that can be empirically set, and $\varepsilon_{i,j}$ is used to overcome the scaling problem. Here we define

$$\varepsilon_{i,j} = \frac{\text{mean}(\rho(x_i, N_i)) + \text{mean}(\rho(x_j, N_j)) + \rho(x_i,x_j)}{3} \tag{4.2}$$

where $\text{mean}(\rho(x_i,N_i))$ is the average value of the distances between x_i and each of its neighbors. μ is recommended to have the value in the range of [0.3, 0.8].

To calculate the fused matrix from multiple types of measurements, we applied a full and sparse kernel on the vertex set V. The full kernel is a normalized weight matrix $\mathbf{P}=\mathbf{D}^{-1}\mathbf{W}$, where $\mathbf{D}$ is the diagonal matrix whose entries $\mathbf{D}(i,i)=\sum_j \mathbf{W}(i,j)$, so that $\sum_j \mathbf{P}(i,j)=1$.

Let N_i represent the set of x_i's neighbors including x_i in G. Given a graph G, KNN is used to measure the local affinity as follows:

$$\boldsymbol{S}(i,j) = \begin{cases} \dfrac{\mathbf{W}(i,j)}{\sum_{k \in N_i} \mathbf{W}(i,k)}, & j \in N_i \\ 0 & \text{otherwise} \end{cases} \tag{4.3}$$

Note that **P** carries the full information about the similarity of each patient to all others, whereas ***S*** only encodes the similarity to the **K** most similar patients for each patient. The algorithm always starts from **P** as the initial state and uses ***S*** as the kernel matrix in the fusion process for both capturing local structure of the graph and computational efficiency.

Given m different data types, we can construct similarity matrices $\mathbf{W}^{(v)}$ using Eq. (4.1) for the vth view, v = 1, 2, ..., m. $\boldsymbol{S}^{(v)}$ are obtained from Eq. (4.3). Below we introduce the network-fusion process, given a set of networks from each individual data.

Let us first consider the case when we have two data types, i.e., $m = 2$. We calculate the status matrices $\mathbf{P}^{(1)}$ and $\mathbf{P}^{(2)}$ from two input similarity matrices; then the kernel matrices $\boldsymbol{S}^{(1)}$ and $\boldsymbol{S}^{(2)}$ are obtained as in Eq. (4.3).

Let $\mathbf{P}_{t=0}^{(1)} = \mathbf{P}^{(1)}$ and $\mathbf{P}_{t=0}^{(2)} = \mathbf{P}^{(2)}$ represent the initial two status matrices at $t = 0$. The key step of SNF is to iteratively update the similarity matrix corresponding to each of the data types as follows:

$$\mathbf{P}_{t+1}^{(1)} = \boldsymbol{S}^{(1)} \times \mathbf{P}_t^{(2)} \times \left(\boldsymbol{S}^{(1)}\right)^T \tag{4.4}$$

$$\mathbf{P}_{t+1}^{(2)} = \boldsymbol{S}^{(2)} \times \mathbf{P}_t^{(1)} \times \left(\boldsymbol{S}^{(2)}\right)^T \tag{4.5}$$

where $\mathbf{P}_{t+1}^{(1)}$ is the status matrix of the first data type after t iterations, whereas $\mathbf{P}_{t+1}^{(2)}$ is the similarity matrix for the second data type. This procedure updates the status matrices each time, generating two parallel interchanging diffusion processes. After t steps, the overall status matrix is computed as follows

$$\mathbf{P}^{(c)} = \frac{\mathbf{P}_t^{(1)} + \mathbf{P}_t^{(2)}}{2} \tag{4.6}$$

The input of SNF algorithm can be feature vectors, pairwise distances, or pairwise similarities. The learned status matrix $\mathbf{P}^{(c)}$ can then be used for clustering and classification. In this work, we mainly focus on clustering and label prediction.

2.2.2 Network Clustering

In the SNFtool package, there is also clustering function employing spectral clustering algorithm. Spectral clustering is effective in capturing global structure of the graph [24].

Given n samples and m measurements, we want to identify C clusters of samples, each of which corresponding to a (known or new) subtype. We associate each sample x_i with a label indicator vector $y_i \in \{0, 1\}$ such that $y_i(k) = 1$ if sample x_i belongs to the kth cluster (subtype), and otherwise $y_i(k) = 0$. So a partition matrix $\mathbf{Y} = \left(y_1^T, y_2^T, \ldots, y_n^T\right)$ is used to represent a clustering scheme.

Given the fused graph, we used spectral clustering to obtain network clusters. It aims to minimize RatioCut [25] by solving the following optimization problem:

$$\begin{gathered} \min_{\mathbf{Q} \in R^{n \times C}} \mathrm{Trace}\left(\mathbf{Q}^T \mathbf{L}^+ \mathbf{Q}\right) \\ s.t.\ \mathbf{Q}^T \mathbf{Q} = 1 \end{gathered} \tag{4.7}$$

where $\mathbf{Q} = \mathbf{Y}(\mathbf{Y}^{\mathrm{T}}\mathbf{Y})^{-1/2}$ is a scaled partition matrix; and $\mathbf{L}^{+}$ denotes the normalized Laplacian matrix $\mathbf{L}^{+} = \mathbf{I} - \mathbf{D}^{-1/2}\mathbf{W}\mathbf{D}^{-1/2}$ given the similarity matrix $\mathbf{W}$. Matrix $\mathbf{D}$ is a network degree matrix with degrees of each node on the diagonal and off-diagonal elements are set to 0.

2.2.3 Normalized Mutual Information

We use normalized mutual information (NMI) to measure the concordance of two clusters. Given two clustering results U and V on a set of data points, NMI is defined as

$$\mathrm{NMI}(U, V) = \frac{I(U, V)}{\sqrt{H(U), H(V)}} \tag{4.8}$$

where $I(U, V)$ is the mutual information between U and V, and $H(U)$ represents the entropy of the clustering U. Details can be found in Ref. [26]. NMI has values between 0 and 1, measuring the concordance of two clustering results. Therefore, a higher NMI refers to higher concordance with the truth, i.e., a more accurate result.

2.2.4 Graph-Based Group Prediction

Besides the subtyping, we can also predict the group label of a new subject based on fused networks. In our work, we used the graph-based learning method, which is a semisupervised method. For the unlabeled data set, we should compute the same types of features as the labeled data set. For each type of data, after normalization, we combined the unlabeled data and labeled data to construct a similarity network between patients. In our work, we constructed three similarity networks from each type of data and then fused two or three of them into one patient—patient network. In a fused network, a test unlabeled subject will be classified into the same group as the labeled one is most similar to the test subject among all of the labeled subjects. The assumption is that the nearby points are likely to have the same label.

3. RESULTS AND DISCUSSIONS

3.1 Method Overview

Given two or more types of data for the same subject/sample (e.g., SZ subjects), we first create a network for each type of data and then fuse these single networks into one similarity network (Figs. 4.1–4.4). The initial

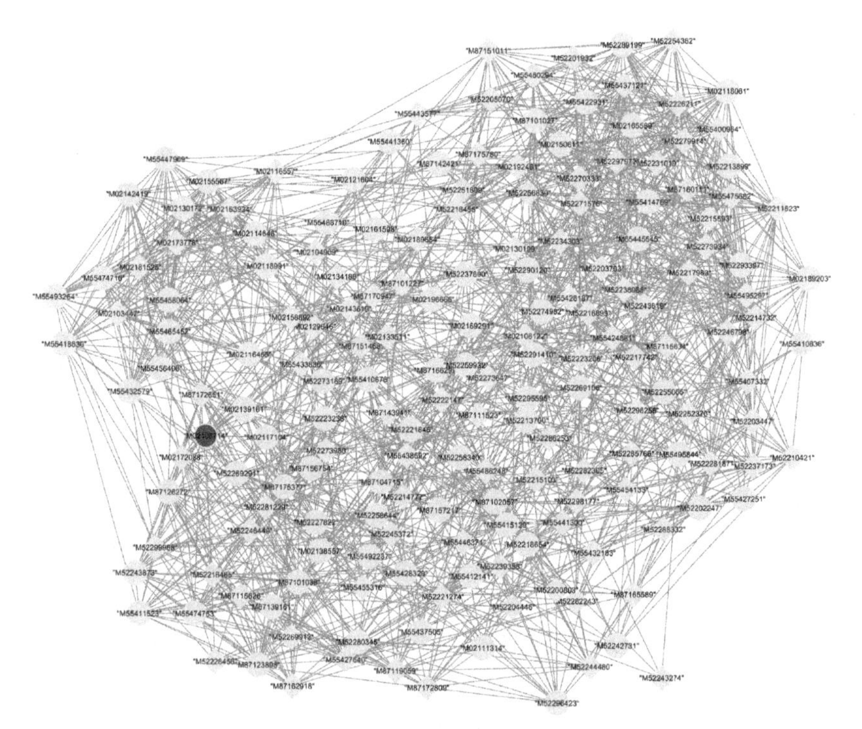

Figure 4.1 A fused network from three data types: single-nucleotide polymorphisms, DNA methylation, and functional magnetic resonance imaging. The *circle* represents disease subject, and the *diamond* represents healthy.

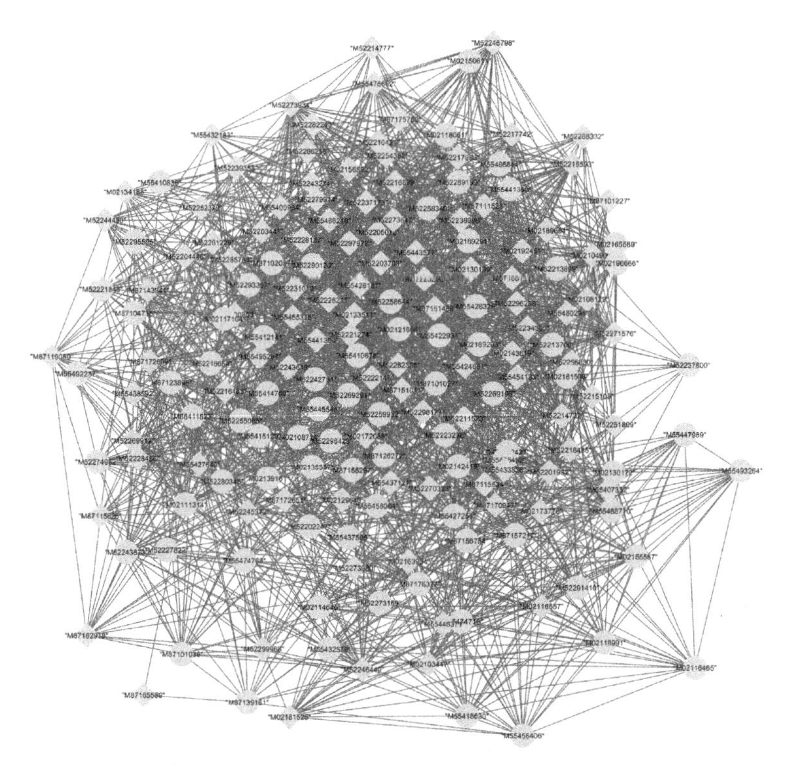

Figure 4.2 A fused network from single-nucleotide polymorphisms and DNA methylation data. The *circle* represents disease subject, and the *diamond* represents healthy.

step is to use a similarity measure for each pair of samples to construct a sample-by-sample similarity matrix for each data type. The matrix represents a similarity network where nodes are samples, and the weighted edges measure the similarity between a pair of samples. The network-fusion step uses a nonlinear method based on message-passing theory [27], which iteratively updates every network, making it more similar to the others in every iteration. After a few iterations, SNF converges to a single network, a common subset whose vertices have strong local affinity.

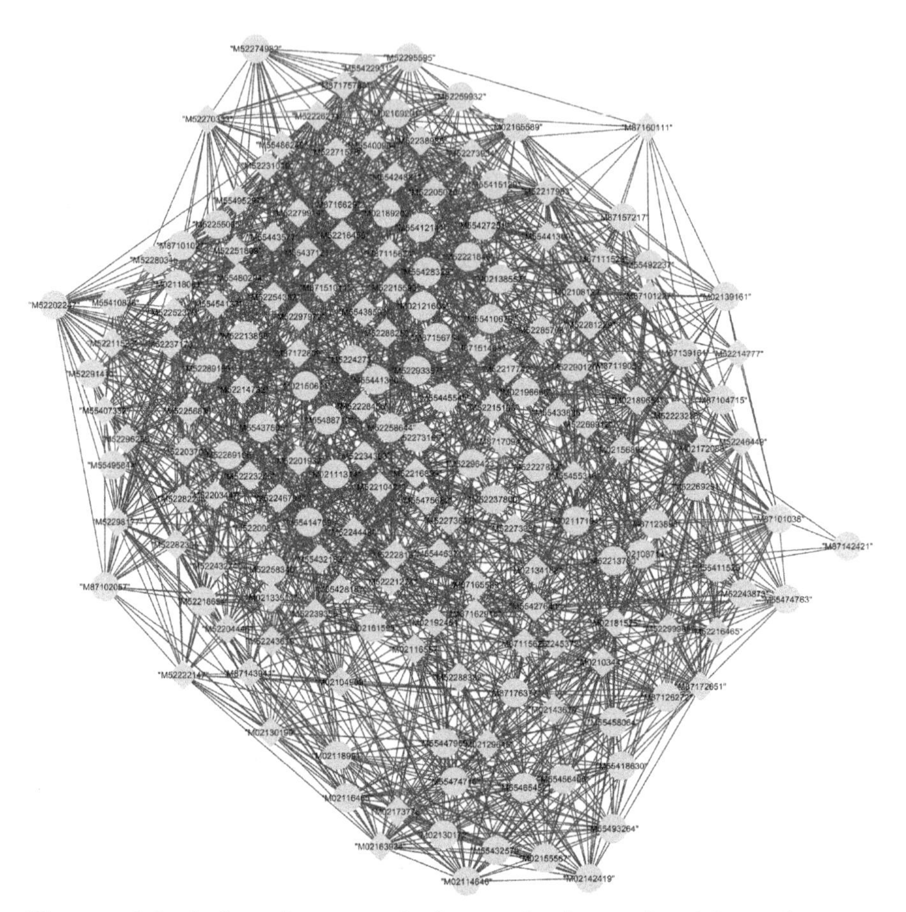

Figure 4.3 A fused network from single-nucleotide polymorphisms and functional magnetic resonance imaging data types. The *circle* represents disease subject, and the *diamond* represents healthy.

3.2 Concordance Between Two Types of Data

Before fusing several single networks into one network, we should calculate the concordance between a pair of data types to measure the compatibility of data sources, which can be characterized via NMI [28]. NMI can help to clarify which data should be combined or not. The greater the

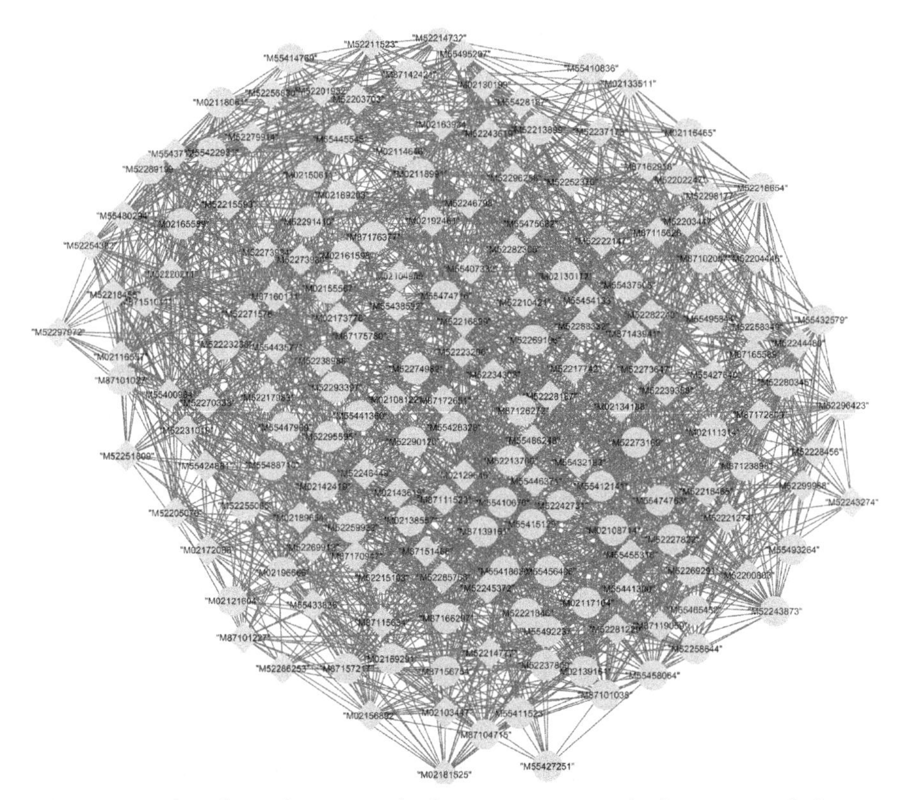

Figure 4.4 A fused network from DNA methylation and functional magnetic resonance imaging data types. The *circle* represents disease subject, and the *diamond* represents healthy.

concordance between the two data types, the more relevant they are. Table 4.1 shows the concordances between each pair of three types of data when the SZ subjects are classified into two groups. C_SM denotes the concordance between

Table 4.1 The concordances between each pair of three types of data to be classified into two groups

	SNPs	Methy	fMRI
SNPs	1	0.006243	0.000114
Methy	0.006243	1	0.014996
fMRI	0.000114	0.014996	1

fMRI, functional magnetic resonance imaging; *SNPs*, single-nucleotide polymorphisms.

SNP data and DNA methylation data. Similarly, C_Sf denotes the concordance between SNP data and fMRI data, C_Mf denotes the concordance between DNA methylation and fMRI data, respectively. From the table, we can see that the rank of their concordances is C_Mf > C_SM > C_Sf.

3.3 Network-Based Subtyping

Based on the fused network, the samples can be classified into healthy and diseased groups using a spectral clustering method. For comparison, we constructed seven networks, including one fused network from all three types of data, three fused networks from each pair of the three types of data, and three single networks from each type of data. We clustered the seven networks separately. For evaluation, we computed the NMI between clusters and the real groups (Table 4.2). Here, the number of clusters is two. If the number of clusters is two, the real subtypes refer to "SUBJECT_RESPONSE_BINARY," i.e., the case and control.

We denote NMI_SMf as the NMI between clusters and real groups using the fused network from SNPs, DNA methylation, and fMRI data. Likewise, NMI_SM, NMI_Sf, NMI_Mf, NMI_SNP, NMI_Methy, and

Table 4.2 The normalized mutual information between clusters and real subtypes for seven sample networks

Network	SMf	SM	Sf	Mf	SNPS	Methy	fMRI
NMI	0.01843	0.04085	0.01850	0.00960	0.03366	0.02047	0.00514

fMRI, network from fMRI data only; *Methy*, network from DNA methylation data only; *Mf*, fused network from DNA methylation and fMRI data; *Sf*, fused network from SNP and fMRI data; *SMf*, fused network from three types of data; *SM*, fused network from SNP and DNA methylation data; *SNP*, network from SNP data only.

NMI_fMRI represent NMI between the predicted clusters and real groups using corresponding networks. For the comparisons of the fused networks, according to the NMI values, the order is NMI_Mf < NMI_SMf < NMI_Sf < NMI_SM. If we compare the fused network with the network from the corresponding single data type, the results are NMI_fMRI < NMI_Sf < NMI_SNP, NMI_SM > NMI_SNP > NMI_Methy, NMI_fMRI < NMI_Mf < NMI_Methy, NMI_fMRI < NMI_SMf < NMI_Methy < NMI_SNP.

From the comparison, it can be seen for SNP and DNA methylation data, their concordance is very small and after two networks being fused the NMIs become greater than that before being fused. Moreover, for the two types of data, after each of them being combined with fMRI data, the NMIs become less than the one based on single type of data, that is, NMI_Sf < NMI_SNP and NMI_Mf < N-MI_Methy. This indicates that SNP and DNA methylation data, both being genetic markers, have more concordance than combined with fMRI imaging at the tissue level.

In addition, it seems that the less the concordance between the two types of data, the more the NMI between the clusters and the real groups will have. It possibly indicates that for classification we should combine those data types with less concordance.

To visualize the clustering of SZ subjects, we present a heatmap of SZ subjects' cluster based on the subjects' illness type (Fig. 4.5). The first more than half of subjects belong to one cluster (the corresponding blue bar), and the latter small part is the other clusters (the corresponding green bar). The above color bar corresponds to the predicted group by spectral clustering method. The group information in the heatmap refers to the groups predicted by hierarchical clustering

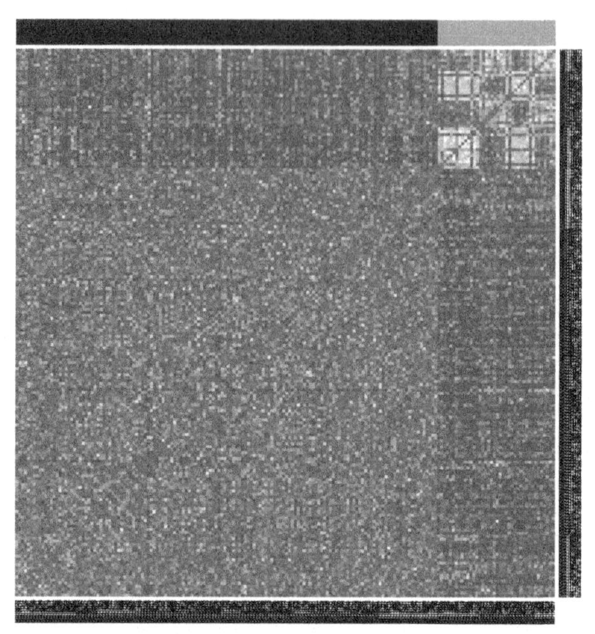

Figure 4.5 The heatmap of schizophrenia (SZ) subjects' clusters. The above *blue bar* (black in print versions) and *green bar* (gray in print versions) refers to two groups predicted by spectral clustering method. The below heatmap is groups predicted by hierarchical clustering method. The labels of x and y axis present the SZ patients' identifications.

method. From the heatmap, it can be seen that the two kinds of groups predicted by two clustering methods are consistent.

3.4 Network-Based Group Prediction

To exploit the potential of using networks as features for diagnosis, we also predicted the groups of new subjects based on network fusion. We used tenfold cross-validation to evaluate the performance of our prediction method. We used acu_SMf to represent the prediction accuracy using the fused network from three types of data. Similar notations are used for other networks. From the

Table 4.3 The prediction accuracy based on different types of data

Network	SMf	SM	Sf	Mf	SNPs	Methy	fMRI
Accuracy	0.64596	0.55657	0.64141	0.61667	0.52879	0.58485	0.54899

fMRI, network from fMRI data only; *Methy*, network from DNA methylation data only; *Mf*, fused network from DNA methylation and fMRI data; *Sf*, fused network from SNP and fMRI data; *SMf*, fused network from three types of data; *SM*, fused network from SNP and DNA methylation data; *SNP*, network from SNP data only.

experimental results (Table 4.3), we can see that based on the fused network from three types of data, the prediction accuracy is the best. For the fused networks from two types of data, the order is acu_Sf > acu_Mf > acu_SM. In addition, it is generally better using the fused network from two types of data than using each single network (acu_Sf > acu_SNP or acu_Sf > acu_fMRI, acu_Mf > acu_Methy or acu_Mf > acu_fMRI, and acu_SM > acu_SNP).

In addition, for the fused networks from pair of data types, the order of the concordances is C_Mf > C_SM > C_Sf. If using spectral clustering method to cluster the networks, the order of NMI between the predicted clusters and the real groups is NMI_Mf < NMI_ Sf < NMI_SM. When predicting the label of unlabeled data by our group prediction method, the order of prediction accuracy is acu_Sf > acu_Mf > acu_SM. It can found that C_Sf is the least, but acu_Sf is the greatest. Does it indicate that for classification we should combine those types of data with low compatibility?

In the current prediction approach, we only achieve a moderate accuracy of classifying SZ based on the fused networks. In the current graph-based prediction approach, we only used the edge weights as the features. We believe that if genomic knowledge is exploited to construct networks and more topological attributes of the fused network are added to the feature sets, the classification accuracy will be improved a lot. It is our future plan. In fact, environment, psychological, and social processes also appear to be important factors for SZ. In addition, the use of some recreational and prescription drugs can cause or worsen symptoms. We will also investigate the associations between SZ and these important factors in the future.

4. CONCLUSIONS

With the rapid development of high-throughput technology, collecting diverse types of genomic data becomes a routine procedure for biological discovery. To make use of the complementary information from different types of data, multitype data integration becomes a pressing issue. In this study, we combined SNP, DNA methylation, and fMRI data based on network fusion method. We created seven networks: three networks for each single data type (SNP, DNA methylation, and fMRI), three fused networks for two data types (SM, Sf, and Mf), and one fused network for all three data types. To test the potential of networks for disease diagnosis, in particular, the added value of using multiomics information, we classified the SZ subjects into two groups based on these seven networks, respectively. We predicted the SZ subjects' group based on the seven networks and used tenfold cross-validation to evaluate the performances. The prediction accuracy based on the fused network for three data types has the highest value among all of seven networks. This further confirms that we should comprehensively utilize multitype data for better diagnosis.

ACKNOWLEDGMENTS

Our work is partially supported by NIH R01 GM109068, R01MH107354, and R01 MH104680.

REFERENCES

[1] M.M. Picchioni, R.M. Murray, Schizophrenia, British Medical Journal 335 (2007) 91–95.
[2] A.P. Association, Diagnostic and Statistical Manual of Mental Disorders, fifth ed., American Psychiatric Publishing, Arlington, 2015.
[3] R. Pies, How "objective" are psychiatric diagnoses? Psychiatry 4 (2007) 18–22.
[4] T. Insel, et al., Research domain criteria (RDoC): toward a new classification framework for research on mental disorders, American Journal of Psychiatry 167 (2010) 748–751.
[5] A.-L. Barabasi, Z.N. Oltvai, Network biology: understanding the cell's functional organization, Nature Reviews Genetics 5 (2004) 101–113.
[6] K. Sun, Uncovering Disease Associations via Biological Network Integration (Ph.D. thesis), Department of Computing, Imperial College London, 2014. Supervised by Dr. Natasa Przulj.
[7] D. Kim, et al., Knowledge boosting: a graph-based integration approach with multi-omics data and genomic knowledge for cancer clinical outcome prediction, Journal of the American Medical Informatics Association 22 (2015) 109–120.
[8] C. Li, M. Liakata, D. Rebholz-Schuhmann, Biological network extraction from scientific literature: state of the art and challenges, Briefings in Bioinformatics 15 (2014) 856–877.
[9] S.W. Chi, J.B. Zang, A. Mele, R.B. Darnell, Argonaute HITS-CLIP decodes microRNA–mRNA interaction maps, Nature 460 (2009) 479–486.
[10] A. Beyer, S. Bandyopadhyay, T. Ideker, Integrating physical and genetic maps: from genomes to interaction networks, Nature Reviews Genetics 8 (2007) 699–710.
[11] B. Zhang, C. Gaiteri, L.G. Bodea, et al., Integrated systems approach identifies genetic nodes and networks in late-onset Alzheimer's disease, Cell 153 (2013) 707–720.

[12] H. Cao, J. Duan, D. Lin, V. Calhoun, Y.-P. Wang, Integrating fMRI and SNP data for biomarker identification for schizophrenia with a sparse representation based variable selection method, BMC Medical Genomics 6 (2013) S2.
[13] H. Cao, S. Li, H.W. Deng, Y.-P. Wang, Identification of genes for complex diseases using integrated analysis of multiple types of genomic data, PLoS One 7 (2012) 1–8.
[14] D. Lin, H. Cao, V.D. Calhoun, Y.-P. Wang, Sparse models for correlative and integrative analysis of imaging and genetic data, Journal of Neuroscience Methods 237 (2014) 69–78.
[15] H. Cao, J. Duan, D. Lin, Y. Shugart, V.D. Calhoun, Y.-P. Wang, Sparse representation based biomarker selection for schizophrenia with integrated analysis of fMRI and SNPs, Neuroimage (2014).
[16] D. Lin, J. Zhang, J. Li, H. Deng, Y. Wang, Integrative analysis of multiple diverse omics datasets by sparse group multitask regression, Frontiers in Cell and Developmental Biology, Section Systems Biology (2014) 62.
[17] R.L. Gollub, et al., The MCIC collection: a shared repository of multi-modal, multi-site brain image data from a clinical investigation of schizophrenia, Neuroinformatics 11 (2013) 367–388.
[18] D. Lin, J. Zhang, J. Li, V. Calhoun, Y.-P. Wang, in: 2013 IEEE 10th International Symposium on Biomedical imaging (ISBI), IEEE, 2013, pp. 278–281.
[19] J. Liu, et al., Methylation patterns in whole blood correlate with symptoms in schizophrenia patients, Schizophrenia Bulletin (2013).
[20] J. Liu, M. Morgan, K. Hutchison, V.D. Calhoun, A Study of the Influence of Sex on Genome Wide Methylation, 2010.
[21] E. Walton, et al., MB-COMT promoter DNA methylation is associated with working-memory processing in schizophrenia patients and healthy controls, Epigenetics 9 (2014) 1101–1107.
[22] B. Wang, A.M. Mezlini, F. Demir, et al., Similarity network fusion for aggregating data types on a genomic scale, Nature Methods 11 (2014) 333–337.
[23] B. Wang, J. Jiang, W. Wang, et al., Unsupervised metric fusion by cross diffusion, in: Proc. IEEE Comput. Soc. Conf. Comput. Vis. Pattern Recognit., 2012, pp. 2997–3004.
[24] U. Luxburg, A tutorial on spectral clustering, Statistics and Computing 17 (2007) 395–416.

[25] Y.C. Wei, C.K. Cheng, Towards efficient hierarchical designs by ratio cut partitioning, in: Proc. Int. Conf. Computer-Aided Design, 1989, pp. 298–301.
[26] N.X. Vinh, J. Epps, J. Bailey, Information theoretic measures for clusterings comparison: variants, properties, normalization and correction for chance, Journal of Machine Learning Research 11 (2010) 2837–2854.
[27] J. Pearl, Probabilistic reasoning in intelligent systems: networks of plausible inference, The Journal of Philosophy 88 (1991) 434–437.
[28] A.S. Ghosh, J. Ghosh, Cluster ensembles—a knowledge reuse framework for combining multiple partitions, Journal of Machine Learning Research 3 (2003) 583–617.

CHAPTER FIVE

Genetic Correlation Between Cortical Gray Matter Thickness and White Matter Connections

Kaikai Shen[1], Vincent Doré[1], Jurgen Fripp[1], Stephen Rose[1], Katie L. McMahon[2], Greig I. de Zubicaray[3], Nicholas G. Martin[4], Paul M. Thompson[5], Margaret J. Wright[4], Olivier Salvado[1]

[1]Australian eHealth Research Centre, CSIRO, Herston, QLD, Australia
[2]University of Queensland, Brisbane, QLD, Australia
[3]Queensland University of Technology (QUT), Brisbane, QLD, Australia
[4]Queensland Institute of Medical Research, Brisbane, QLD, Australia
[5]University of Southern California, Marina del Rey, CA, United States

Contents

Imaging Genetics
ISBN: 978-0-12-813968-4
http://dx.doi.org/10.1016/B978-0-12-813968-4.00005-5

Abstract

The aim of this paper is to investigate the common genetic influence on the cortical thickness (CT) and white matter (WM) connectivity. To this aim, we analyzed cortical gray matter (GM) thickness derived from structural magnetic resonance images and the measures of WM connected to the cortex obtained from diffusion magnetic resonance imaging (dMRI) on a large twin imaging cohort ($N = 308$, average age 22.8 ± 2.3 SD). The CT was estimated using a surface-based approach, and the WM connections were measured based on probabilistic tractography using fiber orientation distributions (FODs) from dMRI. Bivariate genetic modeling was performed on CT and FOD measure of tracts connected to the cortex. Significant genetic correlations between the CT and WM connectivity were found in regions, including the right postcentral gyrus, left posterior cingulated gyrus, right dorsolateral superior frontal gyrus, the right middle, and inferior temporal gyri, suggesting common genetic factors influencing the GM and WM.

Keywords: Cortical thickness; Diffusion MRI; Genetic correlation; Gray matter; Structural MRI; White matter connectivity

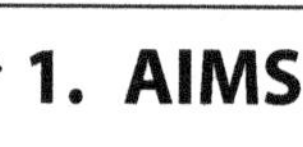

1. AIMS

The cortex covering both hemispheres of the brain is formed by gray matter (GM) in which neuron cell bodies reside. Beneath the cortex, axons form the white matter (WM) pathways communicating with other parts of the central and peripheral nervous system. In the developing brain, the cortical thickness (CT) starts to decrease from childhood and continues throughout adolescence into adulthood [1,2]. This cortical thinning, along with decrease in cortical GM density, is correlated to the WM development [3,4], which has been attributed to cortical area expansion due to WM volume growth [5], synaptic pruning [4,6], and myelin proliferation into the cortex [3,4,7].

The cortical surface area expansion and the cortical thinning appear to be distinct processes as cortical surface area becomes stable from adolescence, whereas cortical thinning progresses linearly into adulthood [8]. In addition, CT and cortical area are also suggested to be under distinct genetic influences as low genetic correlation between them was reported despite that they are both highly heritable [9].

This leads us to the investigations into the link between WM growth and cortical GM thinning, and in particular, the common genetic influence shared by the cortical thinning and the underlying WM. A recent diffusion tensor imaging (DTI) study found the cortical thinning in children and adolescents correlated spatially, especially in sulci, to the decrease in mean diffusivity during WM development [2]. In our study, we used a bivariate genetic model to compute the phenotypic correlation between WM and cortical GM thickness and the genetic correlation between them on a twin cohort. By genetic analysis using twin data, we are able not only to ascertain the correlation between cortical thinning and the WM development but also to test if they may be due to the effects of common genes.

We measured the cortical GM and WM connections using data from two independent modalities of magnetic resonance imaging (MRI): structural MRI (sMRI) for CT and diffusion MRI (dMRI) for WM. Both imaging modalities are able to detect the genetic effects on cortex and WM [10,11]. In this study, we estimated the thickness of cortical GM on sMRI [12] and used a surface-based approach to establish the correspondence of cortical surface across the sample. For WM connections, we carried out a fiber orientation distribution (FOD)–based tractography analysis [13], which we previously employed to find connectivity networks exhibit different degrees of heritability [14]. Region of interest (ROI)–based analysis was used to evaluate the genetic correlation between GM and WM in various parts of the cortex.

2. METHODS

2.1 Subjects and Materials

The twin cohort in our study consisted of 308 subjects (111M, 197F), with average age of 22.8 (±2.3 SD), among which there are 70 pairs of monozygotic twins and 84 pairs of dizygotic twins. The age of the subjects in this cohort ranges from 19 to 29. In this age group, the general cortical thinning continues while the surface area remains stable [8]. A set of 38 subjects (15M, 23F), with average age 23.2 (±2.4 SD, range 20–28) was analyzed to establish the test–retest reliability of the CT. These subjects were scanned twice at on average 3-month intervals (109 ± 51 days, range 36–258 days).

2.2 Image Acquisition

The T1-weighted and diffusion-weighted images were acquired on a 4T Bruker Medspec whole-body scanner (Bruker Medical, Ettingen, Germany). The T1-weighted images were acquired with a magnetization-prepared rapid gradient-echo sequence to resolve anatomy at high resolution. Acquisition parameters were: inversion time (TI)/repetition time (TR)/echo time (TE) = 700/1500/3.35 ms, flip angle = 8 degrees, and slice thickness = 0.9 mm with a $256 \times 256 \times 256$ acquisition matrix. Diffusion images were acquired using a commercial single-shot echo planar multidirection diffusion-weighted sequence, employing a dual bipolar diffusion gradient and a double spin echo.

The imaging parameters were: 55 axial slices; 2 mm slice thickness; field of view 23 × 23 cm; TR/TE 150/92.3 ms; and acquisition matrix 128 × 128, resulting in an in-plane resolution of 1.80 × 1.80 mm. Ninety-four diffusion-weighted images were acquired at $b = 1159\ \mathrm{s/mm^2}$ along with 11 nondiffusion-weighted images ($b = 0$).

2.3 Measuring Cortical Thickness

We used the surface-based approach for CT estimation and analysis [15]. For each subject, the T1-weighted image was segmented into GM, WM, and cerebrospinal fluid using an expectation-maximization (EM) segmentation algorithm [12]. The EM algorithm computed probability maps for each tissue type, which were discretized by assigning each voxel to its most likely tissue type. Partial volume effects due to the limited imaging resolution relative to the size of some anatomical structures were taken into account by the classification and estimation of tissue composition in voxels to increase the precision of CT estimation in regions such as deep sulci. Topological corrections were also applied to deep sulci, and the GM segmentation was constrained to be a continuous layer covering the WM [16]. The segmentation method used nine different atlases and a majority voting rule to reduce the error induced by misregistration. The CT of the resulting GM was computed using a combined voxel-based approach. The CT values were mapped from the image to the cortical surface mesh, which was geometrically smoothed and registered to a common template mesh by a multiscale expectation-maximization iterative closest point (EM-ICP) algorithm [17]. A 10 mm Laplace–Beltrami smoothing was then applied to the CT values on the template mesh. We computed the average CT in each cortical ROI, as defined by automated anatomical labeling atlas [18] on the common template.

2.4 Measuring White Matter Connections

The method for dMRI data processing and analysis was described previously [14]. A brief description follows. The diffusion-weighted images were preprocessed using point-spread function mapping [19], with bias field [20] and motion artifacts corrected [21,22]. Constrained spherical deconvolution [13] was used to estimate the distribution of the fiber population in each voxel, using the response signal from coherently aligned fibers in the corpus callosum.

The dMRI data sets of each subject were warped to an average atlas estimated iteratively. The atlas was created by averaging all the subjects' transformed data [22]. The same transformation field from each subject to the atlas was also used to transform the diffusion tensor map of each subject and to create an average fractional anisotropy (FA) map. By registering the average FA map to the Johns Hopkins University DTI atlas [23] in the standard MNI space, the FOD map of each subject was realigned to the standard space.

The amplitude of FOD peaks was measured on each subject spatially normalized to the MNI space. The three principal FOD peak amplitudes were identified in each voxel of the average FOD template using MRtrix [24], which were then used as template to match the likely peaks in each subject based on angular error. For each voxel, the two largest FOD peaks were used in subsequent analyses described below, unless the second highest FOD peak was lower than 0.1, in which case only one FOD peak was considered.

Whole brain probabilistic fiber tracking [24] was performed on the average FOD template, creating a tractogram of WM networks consisting of cortico—cortical, cortico—subcortical connections, and tracts traveling through the brain stem. We projected the FOD peak amplitude and the test—retest reliability of FOD onto each of the fiber tracts in the tractogram. When a tract intersected voxels with two distinct peaks, the peak along the direction of the tract was chosen. We considered an FOD peak and the passing tract to be in the same direction when they formed an angle less than 45 degrees [14]. In case where the tract passed a voxel where no peak was found to be in the same direction, zero reliabilty or FOD measure was assigned to the tract at that voxel. A reliability mask was used to filter the estimates with test—retest reliability over 0.6. We used the trimmed mean reliability and FOD peak size over the entire stretch to characterize each tract, removing extreme values that arise because of large deviations away from the FOD peak directions or low reliability. The trimmed mean was computed by removing the 5% highest and the 5% lowest values.

We identified in the tractogram the set of tracts that ended in each cortical ROI. For each cortical ROI, the regional average FOD measure was computed by averaging the mean FOD peak size along all the tracts connecting to that cortical region.

2.5 Genetic Correlation Between Connectivity and Cortical Thickness

We evaluated the test—retest reliability of our measures by comparing measurements of the subjects with repeated scans. Intraclass correlation (ICC) [25] was used to assess the test—retest reliability of CT and FOD measures in each ROI. Measurements were adjusted for age and sex.

We computed the genetic correlation between each cortical region's GM thickness and the WM connection's FOD measures to examine the underlying common genetic influences. With the FOD measure computed for each ROI as one variable and the average CT as the other variable, the phenotypic correlation is computed by maximum likelihood estimates. We computed the genetic correlation only for regions with phenotypic correlation greater than 0.1 in absolute value. Given the sample size and the limitation of classical twin study design, the common environment effects were difficult to detect [26]. We therefore used a bivariate AE model (with one additive genetic component A, and one environmental component E) to estimate the genetic correlation. The maximum likelihood estimation and the solution to the bivariate AE model were both computed using OpenMx package [27].

3. RESULTS

The results of test—retest reliability experiment are shown in Table 5.1 in which the ICCs are listed. Most regions have reliable estimates of CT and FOD peak measures, with exceptions of bilateral olfactory cortex, insula, parahippocampal gyrus, amygdala, and the right lingual gyrus, which have reliability ICC of CT estimates below 0.6, and were excluded from the subsequent analyses.

The phenotypic and genetic correlations from the bivariate genetic models are shown in Table 5.2. We computed the genetic correlation for 21 regions with absolute value of phenotypic correlation greater than 0.1 and adjusted the *P*-value of genetic correlation for false discovery rate [28].

Table 5.1 Intraclass correlation showing the test–retest reliability for cortical thickness and fiber orientation distribution (FOD) peak measures by cortical regions

	Cortical thickness		FOD peak size	
	Left	Right	Left	Right
Frontal lobe				
Precentral gyrus	0.776	0.784	0.774	0.791
Superior frontal gyrus dorsolateral	0.788	0.809	0.910	0.896
Superior frontal gyrus orbital part	0.798	0.724	0.884	0.937
Middle frontal gyrus	0.726	0.757	0.935	0.920
Middle frontal gyrus orbital part	0.775	0.815	0.899	0.928
Inferior frontal gyrus opercular part	0.839	0.764	0.913	0.856
Inferior frontal gyrus triangular part	0.758	0.777	0.931	0.945
Inferior frontal gyrus orbital part	0.791	0.778	0.867	0.912
Rolandic operculum	0.731	0.748	0.839	0.818
Supplementary motor area	0.868	0.854	0.841	0.834
Olfactory cortex	0.495	0.522	0.773	0.896
Superior frontal gyrus medial	0.827	0.763	0.957	0.962
Superior frontal gyrus medial orbital	0.773	0.829	0.955	0.940
Gyrus rectus	0.732	0.744	0.857	0.876
Paracentral lobule	0.828	0.856	0.767	0.769
Insula and cingulated gyri				
Insula	0.534	0.573	0.756	0.757
Anterior cingulate and paracingulate gyri	0.770	0.779	0.896	0.921
Median cingulate and paracingulate gyri	0.727	0.756	0.795	0.879
Posterior cingulate gyrus	0.795	0.740	0.800	0.865

(Continued)

Table 5.1 Intraclass correlation showing the test–retest reliability for cortical thickness and fiber orientation distribution (FOD) peak measures by cortical regions—cont'd

	Cortical thickness		FOD peak size	
	Left	**Right**	**Left**	**Right**
Occipital lobe				
Calcarine fissure and surrounding cortex	0.726	0.626	0.825	0.808
Cuneus	0.792	0.650	0.872	0.861
Lingual gyrus	0.687	0.595	0.799	0.801
Superior occipital gyrus	0.852	0.734	0.879	0.817
Middle occipital gyrus	0.844	0.776	0.808	0.796
Inferior occipital gyrus	0.663	0.722	0.745	0.816
Parietal lobe				
Postcentral gyrus	0.799	0.848	0.704	0.762
Superior parietal gyrus	0.885	0.863	0.751	0.799
Inferior parietal gyrus	0.852	0.776	0.731	0.883
Supramarginal gyrus	0.837	0.777	0.679	0.828
Angular gyrus	0.814	0.841	0.841	0.862
Precuneus	0.835	0.790	0.790	0.820
Temporal lobe				
Hippocampus	0.605	0.636	0.725	0.799
Parahippocampal gyrus	0.479	0.522	0.750	0.763
Amygdala	0.380	0.358	0.912	0.840
Fusiform gyrus	0.618	0.640	0.731	0.788
Heschl gyrus	0.743	0.689	0.770	0.731
Superior temporal gyrus	0.753	0.771	0.718	0.809
Temporal pole superior temporal gyrus	0.635	0.487	0.888	0.816
Middle temporal gyrus	0.777	0.760	0.742	0.823
Temporal pole middle temporal gyrus	0.541	0.580	0.716	0.785
Inferior temporal gyrus	0.753	0.726	0.760	0.764

The cortical thickness is averaged in each cortical region, the FOD peak measures were computed by averaging all peak size along all the tracts connecting to that cortical region.

Table 5.2 Phenotypic and genetic correlation between cortical thickness and fiber orientation distribution measure of white matter connections by cortical regions

	Left		Right	
	r_p	r_g	r_p	r_g
Frontal lobe				
Precentral gyrus	0.011		−0.100	
Superior frontal gyrus dorsolateral	−0.014		−0.153	−0.226[a]
Superior frontal gyrus orbital part	−0.136	−0.088	0.048	
Middle frontal gyrus	0.081		−0.149	−0.136
Middle frontal gyrus orbital part	−0.129	−0.088	−0.063	
Inferior frontal gyrus opercular part	0.163	0.283[a]	−0.021	
Inferior frontal gyrus triangular part	0.037		−0.125	−0.057
Inferior frontal gyrus orbital part	−0.100	−0.142	−0.152	−0.170
Rolandic operculum	0.249	0.288[a]	0.050	
Supplementary motor area	0.010		−0.040	
Superior frontal gyrus medial	0.014		0.048	
Superior frontal gyrus medial orbital	−0.097		−0.009	
Gyrus rectus	−0.082		−0.064	
Paracentral lobule	−0.038		−0.033	
Insula and cingulated gyri				
Insula	−0.003		−0.033	
Anterior cingulate and paracingulate gyri	0.022		0.064	
Median cingulate and paracingulate gyri	−0.023		0.091	
Posterior cingulate gyrus	−0.276	−0.430[b]	−0.076	

(Continued)

Table 5.2 Phenotypic and genetic correlation between cortical thickness and fiber orientation distribution measure of white matter connections by cortical regions—cont'd

	Left		Right	
	r_p	r_g	r_p	r_g
Occipital lobe				
Calcarine fissure and surrounding cortex	−0.036		−0.051	
Cuneus	−0.035		0.037	
Lingual gyrus	−0.092			
Superior occipital gyrus	0.045		0.058	
Middle occipital gyrus	−0.022		0.113	0.124
Inferior occipital gyrus	−0.133	−0.138	−0.053	
Parietal lobe				
Postcentral gyrus	−0.117	−0.180	−0.156	−0.311[a]
Superior parietal gyrus	−0.066		−0.058	
Inferior parietal gyrus	−0.040		0.000	
Supramarginal gyrus	−0.036		0.067	
Angular gyrus	−0.045		0.027	
Precuneus	−0.066		−0.011	
Temporal lobe				
Hippocampus	0.073		−0.018	
Fusiform gyrus	−0.202	−0.323[a]	−0.151	−0.133
Heschl gyrus	0.119	0.185	0.041	
Superior temporal gyrus	0.064		−0.019	
Temporal pole superior temporal gyrus	−0.125	−0.155		
Middle temporal gyrus	−0.038		−0.157	−0.320[a]
Inferior temporal gyrus	−0.150	−0.194	−0.219	−0.288[a]

The cortical regions with test−retest ICC > 0.6 in both cortical thickness and FOD measures are used. r_p, phenotypic correlation; r_g, genetic correlation.
[a]FDR adjusted $P < .05$.
[b]FDR adjusted $P < .01$.

The correlations between WM connectivity and GM thickness are overall negative in our cohort, which is consistent with the general trend of cortical thinning and WM development. In particular, in areas such as the right postcentral gyrus, left posterior cingulated gyrus, right dorsolateral superior frontal gyrus, and right middle temporal gyrus, we find significant genetic correlation indicating that the WM and GM traits in these regions may share common genes influencing their development.

4. CONCLUSION

In this paper, we presented an analysis of cerebral cortex using neuroimaging measurements from independent MRI modalities to detect the common genetic influence shared by cortical GM and its WM connections. We measured the CT on sMRI scans and the WM connections by tractographic analysis based on a higher order model of dMRI. We are able to detect genetic correlation between cortical GM thickness and WM connectivity in particular cortical regions including postcentral gyrus, posterior cingulated gyrus, right dorsolateral superior frontal gyrus, and middle temporal gyrus. The genetic correlations in the results suggest the genetic effects on cortical GM thinning via processes related to WM developments such as myelination proliferation.

REFERENCES

[1] P. Shaw, N.J. Kabani, J.P. Lerch, K. Eckstrand, R. Lenroot, N. Gogtay, D. Greenstein, L. Clasen, A. Evans, J.L. Rapoport, J.N. Giedd, S.P. Wise, Neurodevelopmental trajectories of the human cerebral cortex, The Journal of Neuroscience 28 (14) (2008) 3586–3594.

[2] S.N. Vandekar, R.T. Shinohara, A. Raznahan, D.R. Roalf, M. Ross, N. DeLeo, K. Ruparel, R. Verma, D.H. Wolf, R.C. Gur, R.E. Gur, T.D. Satterthwaite, Topologically dissociable patterns of development of the human cerebral cortex, The Journal of Neuroscience 35 (2) (2015) 599–609.
[3] E.R. Sowell, B.S. Peterson, P.M. Thompson, S.E. Welcome, A.L. Henkenius, A.W. Toga, Mapping cortical change across the human life span, Nature Neuroscience 6 (3) (2003) 309–315.
[4] E.R. Sowell, P.M. Thompson, C.M. Leonard, S.E. Welcome, E. Kan, A.W. Toga, Longitudinal mapping of cortical thickness and brain growth in normal children, The Journal of Neuroscience 24 (38) (2004) 8223–8231.
[5] H.L. Seldon, Does brain white matter growth expand the cortex like a balloon? Hypothesis and consequences, Laterality 10 (1) (2005) 81–95.
[6] P.R. Huttenlocher, A.S. Dabholkar, Regional differences in synaptogenesis in human cerebral cortex, The Journal of Comparative Neurology 387 (2) (1997) 167–178.
[7] R.K. Lenroot, N. Gogtay, D.K. Greenstein, E.M. Wells, G.L. Wallace, L.S. Clasen, J.D. Blumenthal, J. Lerch, A.P. Zijdenbos, A.C. Evans, P.M. Thompson, J.N. Giedd, Sexual dimorphism of brain developmental trajectories during childhood and adolescence, NeuroImage 36 (4) (2007) 1065–1073.
[8] I.K. Amlien, A.M. Fjell, C.K. Tamnes, H. Grydeland, S.K. Krogsrud, T.A. Chaplin, M.G.P. Rosa, K.B. Walhovd, Organizing principles of human cortical development – thickness and area from 4 to 30 years: insights from comparative primate neuroanatomy, Cerebral Cortex (2014), http://dx.doi.org/10.1093/cercor/bhu214.
[9] M.S. Panizzon, C. Fennema-Notestine, L.T. Eyler, T.L. Jernigan, E. Prom- Wormley, M. Neale, K. Jacobson, M.J. Lyons, M.D. Grant, C.E. Franz, H. Xian, M. Tsuang, B. Fischl, L. Seidman, A. Dale, W.S. Kremen, Distinct genetic influences on cortical surface area and cortical thickness, Cerebral Cortex 19 (11) (2009) 2728–2735.
[10] G.A.M. Blokland, G.I. de Zubicaray, K.L. McMahon, M.J. Wright, Genetic and environmental influences on neuroimaging phenotypes: a meta-analytical perspective on twin imaging studies, Twin Research and Human Genetics 15 (Special Issue 03) (2012) 351–371.

[11] S.C. Kanchibhotla, K.A. Mather, W. Wen, P.R. Schofield, J.B. Kwok, P.S. Sachdev, Genetics of ageing-related changes in brain white matter integrity – a review, Ageing Research Reviews 12 (1) (2013) 391–401.
[12] O. Acosta, P. Bourgeat, M.A. Zuluaga, J. Fripp, O. Salvado, S. Ourselin, Automated voxel-based 3D cortical thickness measurement in a combined Lagrangian-Eulerian PDE approach using partial volume maps, Medical Image Analysis 13 (5) (2009) 730–743.
[13] J.D. Tournier, C.H. Yeh, F. Calamante, K.H. Cho, A. Connelly, C.P. Lin, Resolving crossing fibres using constrained spherical deconvolution: validation using diffusion-weighted imaging phantom data, NeuroImage 42 (2) (2008) 617–625.
[14] K.K. Shen, S. Rose, J. Fripp, K.L. McMahon, G.I. de Zubicaray, N.G. Martin, P.M. Thompson, M.J. Wright, O. Salvado, Investigating brain connectivity heritability in a twin study using diffusion imaging data, NeuroImage 100 (2014) 628–641.
[15] O. Acosta, J. Fripp, V. Dore, P. Bourgeat, J.M. Favreau, G. Chtelat, A. Rueda, V.L. Villemagne, C. Szoeke, D. Ames, K.A. Ellis, R.N. Martins, C.L. Masters, C.C. Rowe, E. Bonner, F. Gris, D. Xiao, P. Raniga, V. Barra, O. Salvado, Cortical surface mapping using topology correction, partial flattening and 3d shape context-based non-rigid registration for use in quantifying atrophy in Alzheimer's disease, Journal of Neuroscience Methods 205 (1) (2012) 96–109.
[16] A. Rueda, O. Acosta, M. Couprie, P. Bourgeat, J. Fripp, N. Dowson, E. Romero, O. Salvado, Topology-corrected segmentation and local intensity estimates for improved partial volume classification of brain cortex in MRI, Journal of Neuroscience Methods 188 (2) (2010) 305–315.
[17] V. Dore, J. Fripp, P. Bourgeat, K. Shen, O. Salvado, O. Acosta, Surface-based approach using a multi-scale EM-ICP registration for statistical population analysis, in: DICTA 2011, 2011, pp. 13–18.
[18] N. Tzourio-Mazoyer, B. Landeau, D. Papathanassiou, F. Crivello, O. Etard, N. Delcroix, B. Mazoyer, M. Joliot, Automated anatomical labeling of activations in SPM using a macroscopic anatomical parcellation of the MNI MRI single-subject brain, NeuroImage 15 (1) (2002) 273–289.

[19] M. Zaitsev, J. Hennig, O. Speck, Point spread function mapping with parallel imaging techniques and high acceleration factors: fast, robust, and flexible method for echo-planar imaging distortion correction, Magnetic Resonance in Medicine 52 (5) (2004) 1156–1166.
[20] N. Tustison, B. Avants, P. Cook, Y. Zheng, A. Egan, P. Yushkevich, J. Gee, N4itk: improved N3 bias correction, IEEE Transactions on Medical Imaging 29 (6) (2010) 1310–1320.
[21] T. Rohlfing, M.H. Rademacher, A. Pfefferbaum, Volume reconstruction by inverse interpolation: application to interleaved MR motion correction, in: MICCAI 2008, 2008, pp. 798–806.
[22] D. Raffelt, J.D. Tournier, S. Rose, G.R. Ridgway, R. Henderson, S. Crozier, O. Salvado, A. Connelly, Apparent fibre density: a novel measure for the analysis of diffusion-weighted magnetic resonance images, NeuroImage 59 (4) (2012) 3976–3994.
[23] S. Mori, S. Wakana, P.C.M. van Zijl, L.M. Nagae-Poetscher, MRI Atlas of Human White Matter, Elsevier, 2005.
[24] J.D. Tournier, F. Calamante, A. Connelly, MRtrix: diffusion tractography in crossing fiber regions, International Journal of Imaging Systems and Technology 22 (1) (2012) 53–66.
[25] P.E. Shrout, J.L. Fleiss, Intraclass correlations: uses in assessing rater reliability, Psychological Bulletin 86 (2) (1979) 420–428.
[26] P.M. Visscher, S. Gordon, M.C. Neale, Power of the classical twin design revisited: II Detection of common environmental variance, Twin Research and Human Genetics 11 (1) (2008) 48–54.
[27] S. Boker, M. Neale, H. Maes, M. Wilde, M. Spiegel, T. Brick, J. Spies, R. Estabrook, S. Kenny, T. Bates, P. Mehta, J. Fox, OpenMx: an open source extended structural equation modeling framework, Psychometrika 76 (2) (2011) 306–317.
[28] Y. Benjamini, Y. Hochberg, Controlling the false discovery rate: a practical and powerful approach to multiple testing, Journal of the Royal Statistical Society. Series B (Methodological) 57 (1) (1995) 289–300.

CHAPTER SIX

Bootstrapped Sparse Canonical Correlation Analysis: Mining Stable Imaging and Genetic Associations With Implicit Structure Learning

Jingwen Yan[1,2] Lei Du[1], Sungeun Kim[1], Shannon L. Risacher[1], Heng Huang[3], Mark Inlow[4], Jason H. Moore[5], Andrew J. Saykin[1], Li Shen[1,2] , Alzheimer's Disease Neuroimaging Initiative[a]

[1]Indiana University School of Medicine, Indianapolis, IN, United States
[2]Indiana University Indianapolis, Indianapolis, IN, United States
[3]University of Texas at Arlington, Arlington, TX, United States
[4]Rose-Hulman Institute of Technology, Terre Haute, IN, United States
[5]University of Pennsylvania, Philadelphia, PA, United States

[a] Data used in preparation of this article were obtained from the Alzheimer's Disease Neuroimaging Initiative (ADNI) database (adni.loni.usc.edu). As such, the investigators within the ADNI contributed to the design and implementation of ADNI and/or provided data but did not participate in analysis or writing of this report. A complete listing of ADNI investigators can be found at: http://adni.loni.usc.edu/wp-content/uploads/how_to_apply/ADNI_Acknowledgement_List.pdf.

Imaging Genetics
ISBN: 978-0-12-813968-4
http://dx.doi.org/10.1016/B978-0-12-813968-4.00006-7

Contents

Abstract

Sparse canonical correlation analysis (SCCA) based on lasso and structured lasso has been widely studied to explore the complex associations between brain imaging and genetics features. Although those based on lasso have a better control of overall sparsity, they capture only a small portion of signals because of competition within correlated features. Advanced structure-based models provide a partial solution, but final patterns mostly depend on the prior structures applied. In this work, we propose a new framework, bootstrapped sparse canonical correlation analysis (BoSCCA), to explore the stable associations between correlated imaging and genetic data sets and to implicitly reconstruct the hidden structures. We compare the performances of BoSCCA and traditional SCCA using both synthetic and real data. In synthetic data, BoSCCA outperforms traditional SCCA in both association identification and group structure extraction, especially when the signal proportion goes below 5%. In real data, BoSCCA better captures the group structure within regions of interest and linkage disequilibrium blocks among single-nucleotide polymorphisms and yielded more biologically meaningful results.

Keywords: Bootstrap sampling; Feature selection; Imaging genetics; Selection stability; Sparse canonical correlation analysis; Structure learning

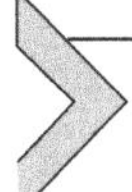

1. INTRODUCTION

Brain imaging genetics study seeks to explore the complex associations between genetic factors such as single-nucleotide polymorphisms (SNPs) and brain quantitative traits (QTs) acquired through multiple neuroimaging techniques. Because of the highly correlated nature of both imaging and genetic data, the power of traditional univariate analyses [1] that examine pairwise correlations become quite limited where features are all treated independently. Sparse canonical correlation analysis (SCCA) [2,3] is a bimultivariate analysis method based on sparsity-induced penalty terms, e.g., Lasso, and has been applied to both real [4] and simulated [5] imaging genetics data, as well as other omics data sets [2,3,6,7]. Although SCCA has demonstrated a better control of correlation problem and also yields very sparse results for easier interpretation, it is only capable of extracting a small portion of correlated features, and this selection procedure is mostly known as to be random.

Various structure-guided SCCA models were proposed recently to address this stability selection issue [3,6,7], where prior structures of data sets were incorporated to guide the learning procedure and to extract correlated features together. But these advanced models only provide partial solutions. Although the resulting patterns mostly depend on the prior structures applied, what structure information is truly functional remains unknown. Thus results based on a randomly selected structure may be questionable. In addition, structure-guided SCCA models are often computationally intensive because of new structure term, and nested cross-validation procedure for parameter tuning makes the computation even more demanding.

To overcome these limitations, in this paper, we propose a new framework BoSCCA, standing for *Bootstrapped SCCA*, to acquire stable imaging and genetic association patterns from high-dimensional correlated data sets. It is particularly worth noticing that the bootstrap not only helps capture those correlated features together but also provides us the insight into the hidden structures that actually function to pull the signals together. We perform an empirical comparison between the proposed bootstrapped canonical correlation analysis (BoCCA) algorithm and a widely used SCCA implementation in the penalized matrix decomposition (PMD) software package (http://cran.r-project.org/web/packages/PMA/) [3] using two synthetic data sets. BoSCCA is observed to have comparable performances with SCCA when the signal proportion of the data is relatively large, e.g., over 5%. But the performance of SCCA declines quickly after the signal proportion goes below 5% whereas BoSCCA still manages to reveal the true patterns. We further test BoSCCA on a real data set to explore the associations between amyloid imaging measures and risk genes of Alzheimer's disease (AD). It is shown that BoSCCA better captures the group structure within regions of interest (ROIs) and linkage disequilibrium (LD) blocks among SNPs and yields biologically meaningful results.

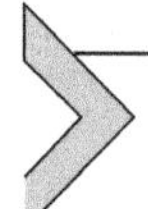

2. BOOTSTRAPPED SPARSE CANONICAL CORRELATION ANALYSIS

Throughout this section, we denote vectors as boldface lowercase letters and matrices as boldface uppercase ones. For a given matrix $\mathbf{M} = (m_{ij})$, we denote its i-th row and j-th column to m_i and m_j, respectively. Let $\mathbf{X} = \{x_1, \ldots, x_n\} \subseteq \Re^p$ be the genotype data (SNP), and

$\mathbf{Y} = \{y_1, \ldots, y_n\} \subseteq \Re^q$ be the imaging QT data, where n, p, and q are the numbers of participants, SNPs, and QTs, respectively.

Canonical correlation analysis (CCA) is a bimultivariate method that explores the linear transformations of variables **X** and **Y** for maximal correlation between **Xu** and **Yv**, which can be formulated as:

$$\max_{\mathbf{u},\mathbf{v}} \sum_{i=1}^{n} \mathbf{u}^T\mathbf{X}^T\mathbf{Y}\mathbf{v} \quad s.t. \; \mathbf{u}^T\mathbf{X}^T\mathbf{X}\mathbf{u} = 1, \; \mathbf{v}^T\mathbf{Y}^T\mathbf{Y}\mathbf{v} = 1 \tag{6.1}$$

where **u** and **v** are the canonical loadings, indicating the significance of each feature in identified associations.

However, the application of traditional CCA in brain imaging genetics studies is very limited because of (1) its requirement of the number of observations n exceeding the combined dimension of **X** and **Y**, and (2) nonsparsity of yielded **u** and **v**, which are difficult to interpret. To address these concerns, SCCA was later proposed where extra penalty terms, $P_1(\mathbf{u}) \leq c_1$ and $P_2(\mathbf{v}) \leq c_2$, were incorporated to impose the sparse constraints. The PMD toolkit [3] provided a widely used SCCA implementation, where the L_1 penalty $P(A) = \sum_{k=1}^{p} |A(k)|$ was used for both P_1 and P_2. Similar to most L_1 penalty-based methodologies, SCCA was found in many studies to identify only a small portion of grouped signals in the associations [7,8], and this selection procedure is usually random because of the competition within correlated signals. Therefore the resulting patterns may vary across experiments and data sets and thus will be hard to interpret. Despite many new models proposed recently to address this issue, most of them are highly dependent on prior structures. Because it is still not clear which structure is actually related with imaging genetic associations, improperly selected structures will possibly bring us biased results.

To overcome this limitation, here we propose a novel framework BoSCCA, combining bootstrap procedure and the traditional SCCA algorithm, to acquire stable imaging and genetic association patterns from high-dimensional correlated data sets. It is particularly worth noticing that the bootstrap procedure not only helps capture those correlated features together but also provides us the insight into the hidden structures that actually function to pull the signals together.

Given the n i.i.d observations $(\mathrm{x}_i,\mathrm{y}_i)$, $i=1, 2, \ldots, n$, represented by matrices $\mathbf{X} = \{\mathrm{x}_1, \ldots, \mathrm{x}_n\} \subseteq \Re^p$ as the imaging QT data and $\mathbf{Y} = \{\mathrm{y}_1, \ldots, \mathrm{y}_n\} \subseteq \Re^q$ as the genotype data (SNP), we perform N bootstrap replications of n data points, where each time $\frac{n}{2}$ observations were independently sampled. To maximize the differences between subsets, we followed [9] to apply a complementary bootstrap strategy, as shown in Fig. 6.1. We first perform only $\frac{N}{2}$ subsample procedures and obtained subsets $\left(\left\langle X_{(1)}, Y_{(1)}\right\rangle, \left\langle X_{(2)}, Y_{(2)}\right\rangle, \ldots, \left\langle X_{\left(\frac{N}{2}\right)}, Y_{\left(\frac{N}{2}\right)}\right\rangle\right)$. Then another $N/2$

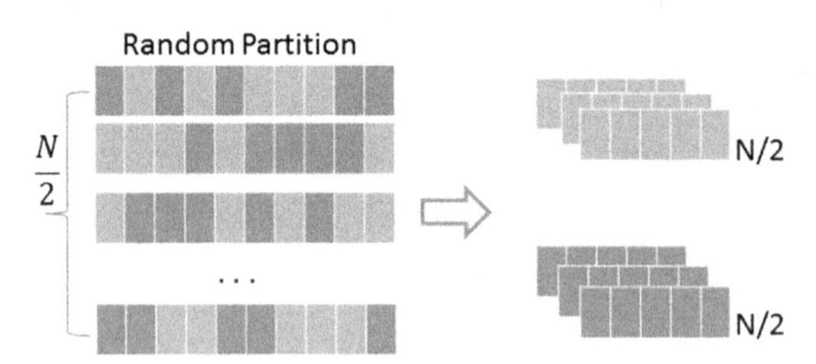

Figure 6.1 Complementary subsampling procedure.

subsets were generated simply by using their complement sets $\left(\left\langle\overline{X_{(1)}},\overline{Y_{(1)}}\right\rangle,\left\langle\overline{X_{(2)}},\ \overline{Y_{(2)}}\right\rangle,\ldots,\left\langle\overline{X_{\left(\frac{N}{2}\right)}},\overline{Y_{\left(\frac{N}{2}\right)}}\right\rangle\right)$. Under each parameter set (c_1, c_2), selection probability of $\mathbf{X}$ and $\mathbf{Y}$ features, $\mathbf{P}_X$ and $\mathbf{P}_Y$, are computed based on these N subsets.

Compared with traditional SCCA, only one more parameter, the bootstrap replication number N, was introduced in BoSCCA. And it is observed later to be very little influential to the final identified patterns as long as it is not too small. Generally it is found that $N > 100$ usually gives similar probability selection patterns. According to Ref. [10], we know that the bootstrap procedure manages to greatly eliminate the effect of parameters on final results. Therefore, there is no need to make the extra effort for parameter tuning. In addition, because the N subsets are independently sampled and include the complement set pairs, concerns regarding the overfitting problems will also be addressed. Shown in Algorithm 6.1 is the detailed algorithm. The final output $\mathbf{P}_X$ and $\mathbf{P}_Y$ are the selection probability matrix, indicating the frequency of selection for imaging and genetic features, respectively, at different parameter settings. Shown in Fig. 6.2 is the derivation procedure of these two matrices. First, given a pair of c_1, c_2 and a subsample set, imaging and genetic features with nonzero canonical loadings are considered as selected and take the value of 1 (Fig. 6.2A). Then, with the same parameter settings, we run the experiment on all N subsample sets. Each entry, with a value between 0 and 1, indicates the frequency each feature gets selected across N bootstrapped experiments. Two three-dimensional selection probability matrices, for imaging and genetic features, respectively, are generated by performing the same analysis across all pairs

Algorithm 6.1 Bootstrapped Sparse Canonical Correlation Analysis

Input:

Normalized $\mathbf{X} = \{x_1, \ldots, x_n\}$, $\mathbf{Y} = \{y_1, \ldots, y_n\}$

$c_1 \in (1, 2, \ldots, \lambda_1)$, $c_2 \in (1, 2, \ldots, \lambda_2)$

Output:

Selection probability of **X** features: $\mathbf{P}_X \in \Re^{N_{c_1} \times N_{c_2} \times p}$

Selection probability of **Y** features: $\mathbf{P}_Y \in \Re^{N_{c_1} \times N_{c_2} \times q}$

1: **for** $j = 1, 2, \ldots, \lambda_1$ **do**
2: **for** $k = 1, 2, \ldots, \lambda_2$ **do**
3: **for** $i = 1, 2, \ldots, \frac{N}{2}$ **do**
4: Generate bootstrap samples $\langle X_{(i)}, Y_{(i)} \rangle$
5: $(\mathbf{u}^i, \mathbf{v}^i) = \text{SCCA}(X_{(i)}, Y_{(i)}, c_1, c_2)$
6: Computer support $\mathbf{S}_\mathbf{u}^i = \{s, u_s^i \neq 0\}$, $\mathbf{S}_\mathbf{v}^i = \{s, \mathbf{v}_s^i \neq 0\}$
7: $\mathbf{P}_X(j, k, \mathbf{S}_\mathbf{u}^i)$, $\mathbf{P}_X(j, k, \mathbf{S}_\mathbf{v}^i)$ increase one
8: Repeat step 4–7 on the complement set $\langle \overline{X_{(i)}}, \overline{Y_{(i)}} \rangle$
9: **end for**
10: **end for**
11: **end for**
12: $\mathbf{P}_X = \mathbf{P}_X/N$, $\mathbf{P}_Y = \mathbf{P}_Y/N$
13: Average $\mathbf{P}_X$ over N_{c_2} and $\mathbf{P}_Y$ over N_{c_1}

of c_1 and c_2 (Fig. 6.2B). Because the selection probability of imaging features is little sensitive to parameter c_2, its corresponding selection matrix is averaged over c_2 with final $\mathbf{P}_X \in \Re^{N_{c_1} x p}$. A similar strategy is also applied on the genetic side to obtain $\mathbf{P}_Y \in \Re^{N_{c_2} x q}$. See Fig. 6.2C.

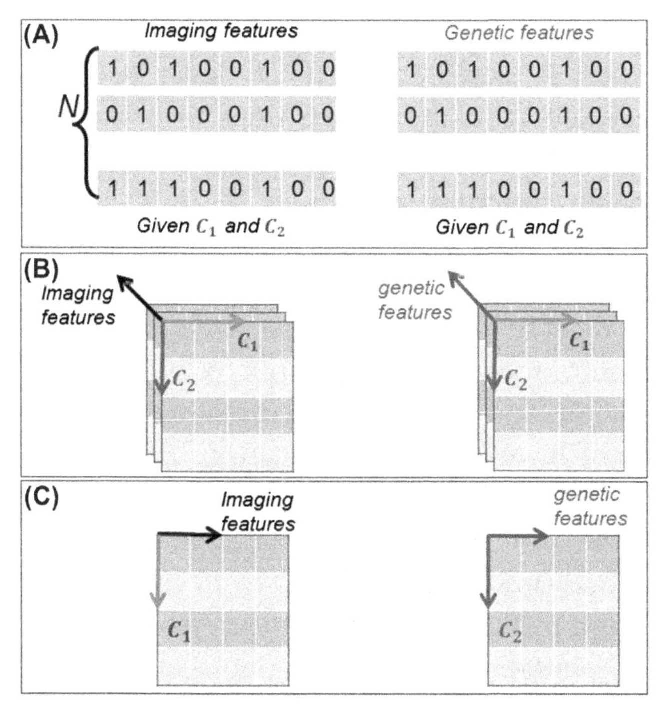

Figure 6.2 Derivation of selection probability matrices $\mathbf{P}_X$ and $\mathbf{P}_Y$. (A) Given a pair of c_1, c_2, imaging and genetic features with value 1 are considered as selected and others as unselected. (B) Given a pair of c_1, c_2, each imaging and genetic feature has a value between 0 and 1 indicating its selection frequency across N bootstrapped experiments. (C) Selection matrix of imaging features is averaged over c_2 with final $\mathbf{P}_X \in \Re^{N_{c_1} \times p}$, and selection matrix of genetic features is averaged over c_1 with final $\mathbf{P}_Y \in \Re^{N_{c_2} \times q}$.

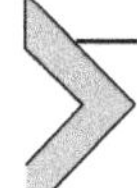

3. EXPERIMENTAL RESULTS

3.1 Results on Synthetic Data

We first performed a comparative study between BoCCA and PMD using synthetic data. We used the following procedure to generate the synthetic data $\mathbf{X}$ and $\mathbf{Y}$ with $n = 800$, $p = 1000$, and $q = 1100$: (1) We created a random positive definite nonoverlapping group structured covariance matrix **M**. (2) Data set **Y** with covariance structure **M** was calculated through Cholesky decomposition. (3) We repeated the above two steps to generate another data set **X**. (4) Canonical loadings **u** and **v** were set based on the group structures of **X** and **Y**, respectively, where all the variables within the group share the same weights. In this initial study, for simplicity, we selected only one group in **Y** to be associated with three groups in **X**. (5) The portion of the specified group in **Y** was replaced based on the **u**, **v**, **X**, and the assigned correlation. We generated eight pairs of **X** and **Y** with correlations ranging from 0.45 to 0.98. The canonical loadings and group structure remained the same across all the synthetic data sets.

We applied BoSCCA and PMD to all eight data sets. The regularization parameters in PMD were chosen to be tuned through permutations. To avoid the overfitting concerns, we performed a fivefold cross-validation experiment for PMD. For BoSCCA, both c_1 and c_2 were set as (1,2,3,4,5). Surprisingly, no signal at all was observed for PMD across all eight data sets. Shown in Fig. 6.3 is the

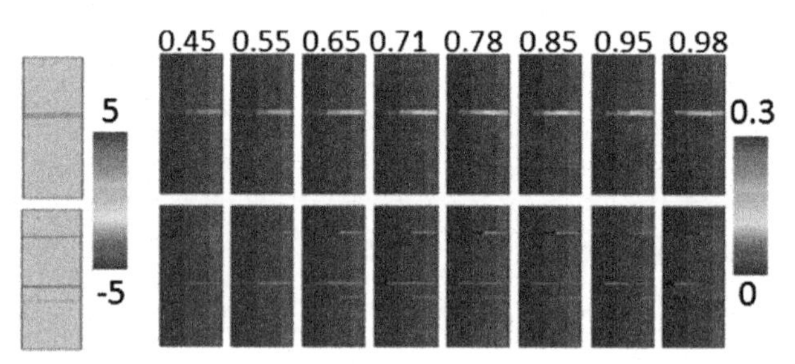

Figure 6.3 Heat maps of selection probability $\mathbf{P}_X^T$ (top blue panels (black in print versions)) and $\mathbf{P}_Y^T$ (bottom blue panels (black in print versions)). Most left: ground truth. Right: heat maps of selection probability. For each blue panel (black in print versions), from left to right indicates different parameters, and from top to bottom indicates features.

selection probability $\mathbf{P}_X$ and $\mathbf{P}_Y$ obtained in BoSCCA. Across all underlying correlations, true signals have much higher probabilities to be selected than background noises, which help themselves to stand out in the selection probability heat map. However, it is worth noticing that the probability is very low, with the maximum only reaching 0.3. And this may be a possible reason that single SCCA could not identify the associations at all.

However, it is still not common for PMD not to identify any signals. Compared with prior studies, the only change in the synthetic data is the signal proportion. Therefore we designed another experiment based on a new synthetic data set ($n = 600$, $p = 1000$, and $q = 1100$) with varied signal proportions (1%–20%). The results are shown in Fig. 6.4. Fig. 6.4A shows the ground truth and Fig. 6.4B

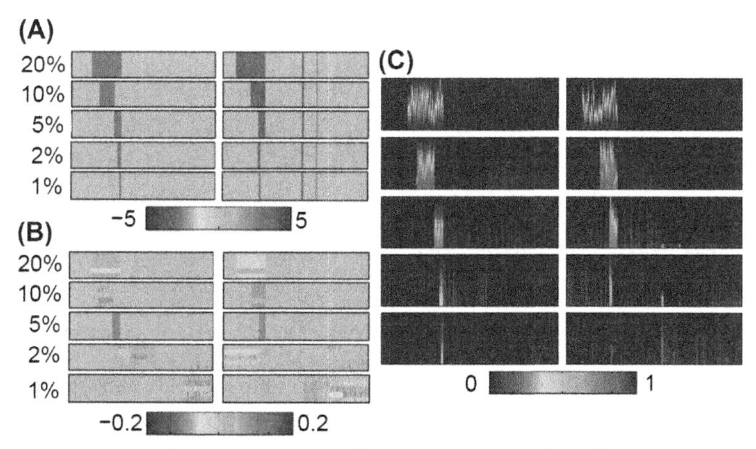

Figure 6.4 Result comparison between penalized matrix decomposition (PMD) and bootstrapped sparse canonical correlation analysis (BoSCCA) on synthetic data sets with varied signal proportions. (A) Ground truth; (B) PMD result: fivefold training weights of **X** and **Y** features. (C) BoSCCA result: Heat maps of selection probability $\mathbf{P}_X$ (left blue panels (black in print versions)) and $\mathbf{P}_Y$ (right blue panels (black in print versions)). For each blue panel (black in print versions), from left to right indicates features, and from top to bottom indicates different parameters.

and C are PMD results and BoSCCA results, respectively. It is clear that PMD functions very well when the signal proportion is greater than 5%, though there are still some minor signals missed. But when the signal proportion goes below 5%, its performance declines quickly and completely fails after signals reaching 2%. For BoSCCA, true signals get selected in the order of their importance in the ground-truth associations when the parameters c_1 and c_2 gradually increase. When the signal proportion is above 5%, the true signals usually all have very high selection probabilities. After signal proportion decreases below that threshold, the selection probabilities of true signals decrease accordingly. But still the patterns can be easily captured on the probability heat maps.

3.2 Results on Brain-Wide Neuroimaging Genetics Data

Brain imaging and genetics data are generally very high dimensional, but true associations usually only involve a small number of brain regions and a very small portion of the genome. Thus, applying traditional SCCA might lead to biased patterns based on our earlier findings. Although targeted analyses may be a potential solution, it narrows down the search space greatly and thus may only yield sub-optimal results. Here, we apply BoSCCA algorithm, to an amyloid imaging genetic analysis in the study of AD.

BoSCCA was empirically evaluated using the amyloid imaging and genotyping data obtained from the Alzheimer's Disease Neuroimaging Initiative (ADNI) database (adni.loni.usc.edu). One goal of ADNI has been to test whether serial magnetic resonance imaging, positron emission tomography, other biological markers, and clinical and neuropsychological assessment can be combined to measure the progression of mild cognitive impairment (MCI) and early AD. For up-to-date information, see www.adni-info.org.

Amyloid imaging measures were obtained following the pipeline in Ref. [11]. Instead of ROI measures, we down-sampled the brain (factor = 2) to obtain 23,133 voxel-level amyloid measurements. Genotype data of both ADNI-1 and ADNI-2/GO phases were obtained from Laboratory of Neuro Imaging (LONI) (adni.loni.usc.edu). All the SNPs of 22 AD risk genes [12] (boundary: gene ±100 kb) were extracted based on the quality controlled and imputed data combining two phases together. Only SNPs available in Illumina 610Quad and/or OmniExpress arrays were included in the analysis. As a result, we had 7517 SNPs included. In total, we have 980 non-Hispanic Caucasian participants with both complete amyloid measurements and genotyping data, including 231 healthy control (HC), 291 early MCI (eMCI), 192 late MCI (LMCI), 195 AD, and 91 subjective memory complaint (SMC) participants (Table 6.1). Using the regression weights derived from the HC participants, amyloid imaging measures were preadjusted to remove the effects of baseline age, gender, and education.

Shown in Fig. 6.5 are the imaging and genetic markers identified in the first pair of canonical components, where $(c_1, c_2) = (17, 3)$ corresponds to the maximum averaged correlation coefficient 0.725 across $N = 200$ bootstrap replications. In Fig. 6.5B, frontal medial orbital gyrus, anterior cingulate gyrus, and posterior cingulate gyrus show very high selection probabilities. Other brain regions, such as olfactory, middle frontal gyrus, putamen, and precuneus, also demonstrate their potential to be affected with slightly smaller probabilities. Amyloid deposition of most of these regions has been previously reported to relate with AD but few of them have managed to extract them together.

Table 6.1 Participant characteristics

	HC	eMCI	LMCI	AD	SMC
Number	231	291	192	175	91
Gender (M/F)	119/112	166/125	110/82	105/70	37/54
Age (mean ± std)	74.17 ± 5.79	71.25 ± 7.20	72.12 ± 7.97	73.69 ± 7.77	71.8 ± 5.73
Education (mean ± std)	16.43 ± 2.67	16.12 ± 2.64	16.32 ± 2.78	15.87 ± 2.76	16.81 ± 2.59

AD, Alzheimer's disease; *eMCI*, early mild cognitive impairment; *HC*, healthy control; *LMCI*, late mild cognitive impairment; *SMC*, subjective memory complaint.

For the gene side, all the significant genetic markers were from APOE and its neighbors APOC1 and TOMM40, which are well-known AD risk regions. This gives us the evidence that BoSCCA can not only identify the true signals but also manage to reveal some hidden structures. Another point worth mentioning is that although all SNPs around APOE are highly correlated, BoSCCA does not extract them all without any differentiation.

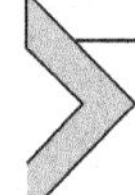

4. CONCLUSIONS

We proposed a new framework BoSCCA, combining the bootstrap procedure and traditional SCCA algorithm, to address the stability selection issue of SCCA and to eliminate the constraints introduced by prior structures while still preserving the capability of identifying correlated features together. When comparing BoSCCA with a widely used version of SCCA on synthetic data, it is observed that the performance of SCCA declines quickly after the signal

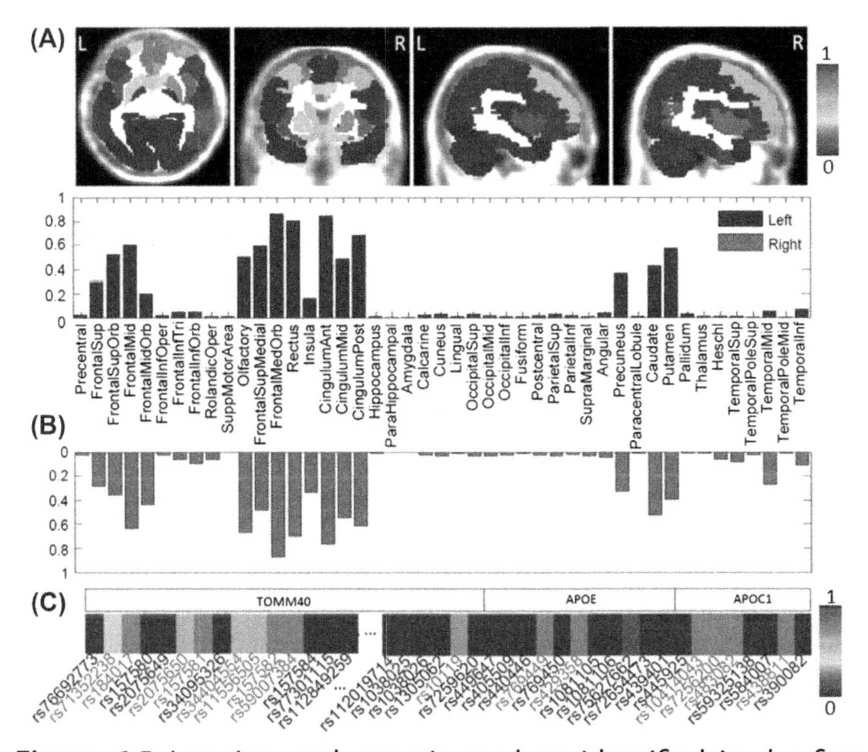

Figure 6.5 Imaging and genetic markers identified in the first pair of canonical components: (A) selection probability on brain map, (B) selection probability in histogram, and (C) selection probability of top single-nucleotide polymorphisms. For (A) and (B), the selection probability of each region of interest is represented by its top 10%.

proportion goes below 5%. On the contrary, BoSCCA can still capture most signals and reveal the hidden structures as well. In real data set, BoSCCA manages to restore the group structure within ROIs and LD blocks between SNPs and yielded biologically meaningful results. Currently, this framework is still limited as a semiautomated method, where signals are manually extracted from the probability heat maps. Further efforts will be made to extend this framework into a fully automated method.

ACKNOWLEDGMENTS

At Indiana University, this work was supported by NIH R01 LM011360, U01 AG024904, RC2 AG036535, R01 AG19771, P30 AG10133, UL1 TR001108, R01 AG 042437, and R01 AG046171; NSF IIS-1117335; DOD W81XWH-14-2-0151, W81XWH-13-1-0259, and W81XWH-12-2-0012; NCAA 14132004; and CTSI SPARC Program. At the University of Texas at Arlington, this work was supported by NSF IIS-1117965, IIS-1302675, IIS-1344152, and DBI-1356628. At the University of Pennsylvania, this work was supported by NIH R01 LM011360, R01 LM009012, and R01 LM010098.

Data collection and sharing for this project was funded by the Alzheimer's Disease Neuroimaging Initiative (ADNI) (National Institutes of Health Grant U01 AG024904) and DOD ADNI (Department of Defense award number W81XWH-12-2-0012). ADNI is funded by the National Institute on Aging, the National Institute of Biomedical Imaging and Bioengineering, and through generous contributions from the following: AbbVie, Alzheimers Association; Alzheimers Drug Discovery Foundation; Araclon Biotech; BioClinica, Inc.; Biogen; Bristol-Myers Squibb Company; CereSpir, Inc.; Eisai Inc.; Elan Pharmaceuticals, Inc.; Eli Lilly and Company; EuroImmun; F. Hoffmann-La Roche Ltd and its affiliated company Genentech, Inc.; Fujirebio; GE Healthcare; IXICO Ltd.; Janssen Alzheimer Immunotherapy Research & Development, LLC.; Johnson & Johnson Pharmaceutical Research & Development LLC.; Lumosity; Lundbeck; Merck & Co., Inc.; Meso Scale Diagnostics, LLC.; NeuroRx Research; Neurotrack Technologies; Novartis Pharmaceuticals Corporation; Pfizer Inc.; Piramal Imaging; Servier; Takeda Pharmaceutical Company; and Transition Therapeutics. The Canadian Institutes of Health Research is providing funds to support ADNI clinical sites in Canada. Private sector contributions are facilitated by the Foundation for the National Institutes of Health (www.fnih.org). The grantee organization is the Northern California Institute for Research and Education, and the study is coordinated by the Alzheimer's Disease Cooperative Study at the University of California, San Diego. ADNI data are disseminated by the Laboratory for Neuro Imaging at the University of Southern California.

REFERENCES

[1] L. Shen, S. Kim, et al., Whole genome association study of brain-wide imaging phenotypes for identifying quantitative trait loci in MCI and AD: a study of the ADNI cohort, Neuroimage 53 (3) (2010) 1051–1063.
[2] E. Parkhomenko, D. Tritchler, J. Beyene, Sparse canonical correlation analysis with application to genomic data integration, Statistical Applications in Genetics and Molecular Biology 8 (2009) 1–34.
[3] D.M. Witten, R. Tibshirani, T. Hastie, A penalized matrix decomposition, with applications to sparse principal components and canonical correlation analysis, Biostatistics 10 (3) (2009) 515–534.
[4] D. Lin, V.D. Calhoun, Y.P. Wang, Correspondence between fMRI and SNP data by group sparse canonical correlation analysis, Medical Image Analysis 18 (6) (2013) 891–902.
[5] E. Chi, G. Allen, et al., Imaging genetics via sparse canonical correlation analysis, in: 2013 IEEE 10th Int. Symp. on Biomedical Imaging (ISBI), 2013, pp. 740–743.
[6] J. Chen, F.D. Bushman, et al., Structure-constrained sparse canonical correlation analysis with an application to microbiome data analysis, Biostatistics 14 (2) (2013) 244–258.
[7] X. Chen, H. Liu, J.G. Carbonell, Structured sparse canonical correlation analysis, in: International Conference on Artificial Intelligence and Statistics, 2012.
[8] L. Du, J. Yan, et al., A novel structure-aware sparse learning algorithm for brain imaging genetics, in: Medical Image Computing and Computer-Assisted Intervention (MICCAI), 2014, pp. 329–336.
[9] R.D. Shah, R.J. Samworth, Variable selection with error control: another look at stability selection, Journal of the Royal Statistical Society: Series B (Statistical Methodology) 75 (1) (2013) 55–80.
[10] N. Meinshausen, P. Bhlmann, Stability selection, Journal of the Royal Statistical Society: Series B (Statistical Methodology) 72 (4) (2010) 417–473.
[11] J. Yan, L. Du, et al., Transcriptome-guided amyloid imaging genetic analysis via a novel structured sparse learning algorithm, Bioinformatics 30 (17) (2014) 564–571.
[12] J.C. Lambert, I.V. Carla, et al., Variable selection with error control: another look at stability selection, Nature Genetics 45 (12) (2014) 1452–1458.

CHAPTER SEVEN

A Network-Based Framework for Mining High-Level Imaging Genetic Associations

Hong Liang[1,4] Xianglian Meng[1,3] Feng Chen[1], Qiushi Zhang[1,2] Jingwen Yan[4,5] Xiaohui Yao[4,5] Sungeun Kim[4], Lei Wang[1], Weixing Feng[1], Andrew J. Saykin[3], Jin Li[1,a], Li Shen[4,a] , Alzheimer's Disease Neuroimaging Initiative[b]

[1]Harbin Engineering University, Harbin, China
[2]Northeast Dianli University, Jilin, China
[3]Habin Huade University, Harbin, China
[4]Indiana University School of Medicine, Indianapolis, IN, United States
[5]Indiana University School of Informatics and Computing, Indianapolis, IN, United States

[a] Supported by grants from National Key Scientific Instrument and Equipment Development Projects of China (2012YQ04014010), National Natural Science Foundation of China (61471139), International Science & Technology Cooperation Program of China (2014DFA70470) and Natural Science Foundation of Heilongjiang (F201241); and by NIH R01 LM011360, U01 AG024904, RC2 AG036535, R01 AG19771, P30 AG10133, UL1 TR001108, NSF IIS-1117335, DOD W81XWH-14-2-0151, and NCAA 14132004 at IU.

[b] Data used in preparation of this article were obtained from the Alzheimer's Disease Neuroimaging Initiative (ADNI) database (adni.loni.usc.edu). As such, the investigators within the ADNI contributed to the design and implementation of ADNI and/or provided data but did not participate in analysis or writing of this report. A complete listing of ADNI investigators can be found at: http://adni.loni.usc.edu/wp-content/uploads/how_to_apply/ADNI_Acknowledgement_List.pdf.

Imaging Genetics
ISBN: 978-0-12-813968-4
http://dx.doi.org/10.1016/B978-0-12-813968-4.00007-9

Contents

Abstract

Genome-wide association studies (GWASs) have identified many individual genes associated with brain imaging quantitative traits (QTs) in Alzheimer's disease (AD). However, single marker level association may not be able to address the underlying biological interactions associated with disease mechanism. In this paper, we propose a network-based framework, guided by protein–protein interaction data, for mining high-level imaging genetic associations. Multilevel GWASs are conducted to investigate the genetic main effect on subcortical imaging measures. Network interface miner for multigenic interactions is employed to discover the trait prioritized subnetworks that are significantly associated with a trait. For several identified significant QT-subnetwork associations, we map the QTs to the imaging domain and perform functional annotation and network analysis for the genes in the subnetwork. The gene-level GWAS yielded significant hits within the *APOE* and *APOC1* regions, which were previously implicated in AD. Pathway analysis was performed to make functional annotation for a few identified subnetworks and discovered several pathways related to degenerative diseases. The imaging results revealed that significant effects emerged on subcortical regions especially on hippocampus.

Keywords: Imaging genetics; Network-based analysis; Pathway analysis; Subcortical quantitative trait

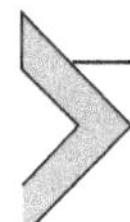

1. INTRODUCTION

Highly complex brain structure and function are strongly affected by genetic factors in neurodegenerative disorder [1,2]. Genome-wide association studies (GWASs) have greatly facilitated the identification of genetic markers associated with brain imaging quantitative traits (QTs) in Alzheimer's disease (AD) [1,3–5]. However, the genetic variation explained by GWAS suffers from a lack of sufficient statistical power to identify markers with small individual effect sizes. It ignores that the potential for gene–gene interaction has been proposed to be one of possible reasons for the so-called "missing heritability" of GWAS, along with other possible factors such as rare variants and environmental factors. So some studies explore the gene–gene interaction, rare variants, and environmental factors in genetic association testing for AD [6–14]. In addition, biological interactions may be very important in contributing a collective effect on disease outcomes [15]. Thus gene set enrichment analysis is employed to identify pathways associated with AD in recent studies [16–18]. However, pathway analysis has a few limitations. For example, varieties with less marginal association can be missed. Overlapping genes among multiple pathways are sometimes ignored. Recently, as an alternative strategy, network-based analysis guided by biologically relevant connections from public databases has attracted attention in genetic association studies.

Given that the etiology of AD might depend on functional protein–protein interaction (PPI) networks, we propose a bioinformatics framework using a PPI network as background knowledge to search for high-level genetic associations with subcortical imaging measures. The proposed framework is shown in Fig. 7.1. First, single marker main effects are examined at the single-nucleotide polymorphism (SNP) level and gene level, respectively. Second, the gene-level *P*-values are used to mine the trait prioritized subnetworks. Then, for the gene set within each significant subnetwork, we perform functional annotation and imaging annotation subsequently. Finally, for each significant subnetwork, the overlapping analysis is conducted to find

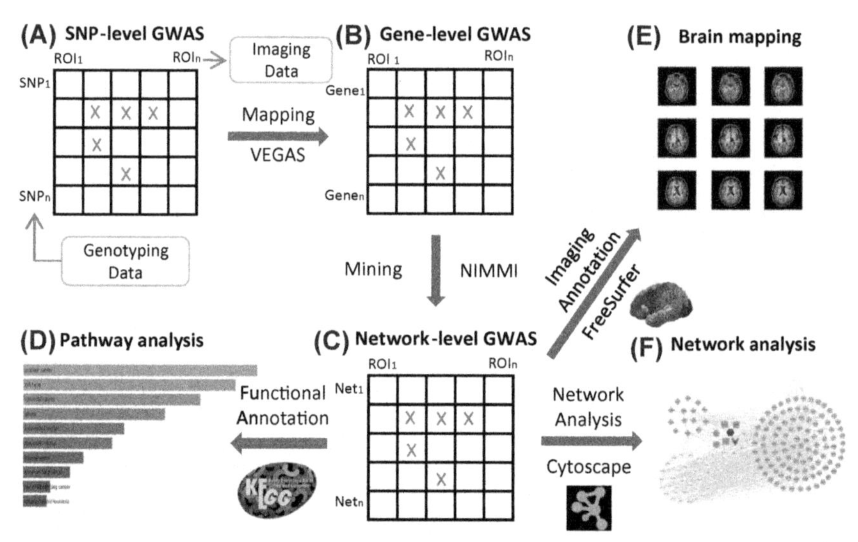

Figure 7.1 An overview of the proposed analysis framework. (A) Perform SNP-based genome-wide association study (GWAS) of subcortical imaging measures. (B) Map SNPs to genes using Versatile Gene-based Association Study (VEGAS). (C) Mine "trait prioritized subnetworks" with network interface miner for multigenic interactions (NIMMI). (D) Perform KEGG pathway analysis for each network-based gene set. (E) Map the subnetworkwise *P*-value to brain. (F) Do network analysis using Cytoscape. *SNP*, single-nucleotide polymorphism.

the highly overrepresented genes associated with most of subcortical measures and visualize them.

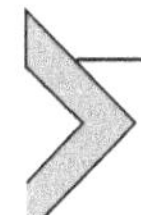

2. METHODS AND MATERIALS

2.1 Subjects and Data

Data used in the preparation of this article were obtained from the Alzheimer's Disease Neuroimaging Initiative (ADNI) database (adni.loni.usc.edu). The ADNI was launched in 2003 as a public—private partnership, led by Principal Investigator Michael W. Weiner, MD. The primary goal of ADNI has been to test whether serial magnetic resonance imaging (MRI), positron emission tomography, other biological markers, and clinical and neuropsychological assessment can be combined to measure the progression of mild cognitive impairment (MCI) and early AD. For up-to-date information, see www.adni-info.org.

Baseline 3T MRI scans data were run with FreeSurfer version 5.1 using cross-sectional processing, demographic information, and baseline diagnosis for all the ADNI-1 and ADNI-GO/2 phases were downloaded [19]. We focused our study on examining the volume measures of the 14 subcortical regions of interest (ROIs) (shown in Table 7.1) in both hemispheres.

Genotype data of all ADNI-1 and ADNI-GO/2 phases were downloaded, and then quality controlled and combined as described in Ref. [8]. A total of 866 non-Hispanic Caucasian participants with both complete subcortical imaging measurements and genotyping data were included in the study. The study sample (N = 866) included 183 cognitively normal, 95 significant memory concerns, 281 early MCI, 177 late MCI, and AD 130 subjects.

Table 7.1 14 FreeSurfer subcortical regions of interest

Measure	Description	Region
AmygVol	Amygdala volume	Subcortical (temporal)
HippVol	Hippocampus volume	
AccumVol	Accumbens volume	Subcortical (striatum/basal ganglia)
CaudVol	Caudate volume	
PallVol	Pallidum volume	
PutamVol	Putamen volume	
ThalVol	Thalamus volume	Subcortical (thalamus)

2.2 Multilevel Genome-Wide Association Studies and Trait Prioritized Subnetwork Identification

GWAS was performed to examine the main effects of 563,980 SNPs on 14 subcortical measures as QTs. Linear regression model was performed using PLINK to determine the each pair of SNP-QT association (http://pngu.mgh.harvard.edu/purcell/plink/) [20]. An additive genetic model was tested with age, gender, and brain volume as covariates.

We employed Versatile Gene-based Association Study (VEGAS) to assign all 563,980 SNPs to their respective genes and calculate an empirical gene-based *P*-value using Monte Carlo simulations (http://gump.qimr.edu.au/VEGAS/), and for each trait around 23,900 genes were obtained [21]. Any SNP that falls within a 10 kB flanking region of a gene was assigned to the gene. The linkage disequilibrium for each gene is estimated using HapMap CEU populations as a reference dataset.

We used a tool named NIMMI to identify and prioritize the subnetworks for each subcortical trait [22].

It constructed a PPI network from BioGrid database and created 2849 subnetworks by performing an exhaustive search consisting of paths of length 2 from a starting node. A modified PageRank algorithm was employed to calculate the weights of proteins in each subnetwork [23]. These weights were then combined with gene-based *P*-values from VEGAS that represented its network centrality. Thus the prioritized subnetworks associated with each subcortical were identified to facilitate the downstream analysis. For each subnetwork, the following information was collected: names of significant and nonsignificant genes, number of significant and nonsignificant genes, total number of genes in a network, network number, combined Z-score, network *P*-value, and corrected *P*-value.

2.3 Pathway and Network Analysis

Subcortical brain regions may jointly lead to abnormal behavior and disease such as AD [24]. So we hypothesize that, if a particular subnetwork has significant network-level *P*-value across all 14 subcortical QTs, it may be likely to be associated with certain biological function and serve as a valuable candidate subnetwork for further analysis. For these identified significant subnetworks, we evaluate them for pathway enrichment analysis, imaging annotation, and network overlapping genes analysis.

To determine the functional relevance of the enriched subnetworks, we also tested whether genes from each subnetwork were overrepresented for specific neurobiological functions, signaling pathways, or complex neurodegenerative diseases. Genes from a significant subnetwork were entered into the functional annotation tool, Enrichr [25] to perform pathway enrichment analysis using pathway database KEGG. *P*-values of subnetworks were mapped onto brain ROIs. We employed Cytoscape (http://www.cytoscape.org/index.html) to plot the identified subnetworks, where each gene was labeled by the number of its significantly associated QTs.

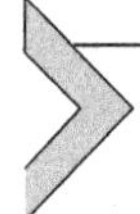

3. RESULTS AND DISCUSSIONS

3.1 Significant Findings From Multilevel Association Results

SNP-level GWAS was performed using PLINK on each QT. Using Bonferroni corrected *P*-value of .05 as the threshold, significant associations were identified between loci on chromosome 19 and a few subcortical measures. Top two SNPs, rs769449 from *APOE* and rs4420638 from *APOC1*, have the strongest associations with left and right hippocampus as shown in Fig. 7.2.

Gene-level association *P*-values were computed based on SNP-level *P*-values by VEGAS. Using Bonferroni corrected *P*-value of .05 as the threshold, APOC1 and TOMM40 have significant association with all the 14 subcortical measures, including left hippocampus, right hippocampus, and right amygdala, which were previously demonstrated to have relation with AD [26]. Fig. 7.3 shows an example Manhattan plot of gene-based GWAS of right hippocampal volume.

Using Bonferroni corrected *P*-value of 0.05 as the threshold, 700 subnetworks were identified by NIMMI to be involved in at least one significant QT-subnetwork association. Fig. 7.4 shows an example heat map of QT-subnetwork associations, where uncorrected $P < 1e\text{-}13$ was used to label the "significant" entries. Three subnetworks, 1431 (146 genes), 2778 (151 genes), and 6682 (157 genes) are significantly associated with all 14 QTs.

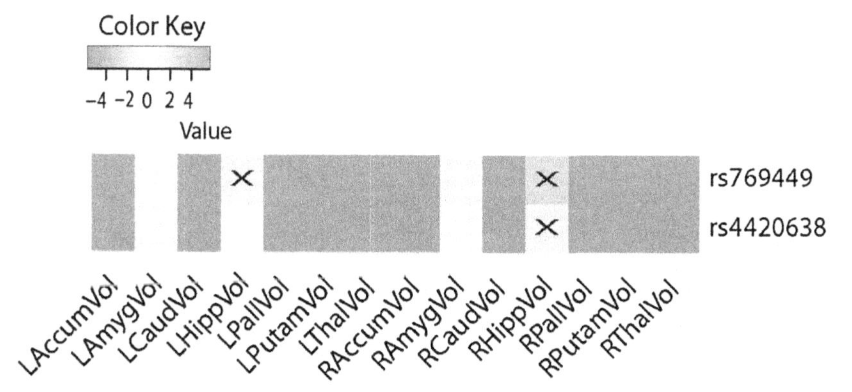

Figure 7.2 Heat map of SNP-based genome-wide association study of subcortical quantitative traits (QTs) at the significance level of $P < 8.87\text{e-}8$ (corrected for 563,980). $-\log_{10}(P\text{-values})$ from each association are mapped and displayed in the heat map. Heat map blocks labeled with "x" reach the significance level of $P < 8.87\text{e-}8$. SNPs and QTs with at least one significant finding are included in the heat map, and thus each row or column contains at least one "x". *SNP*, single-nucleotide polymorphism.

To further illustrate the high-level imaging genetic associations, we map *P*-values of these subnetworks onto the brain, see Fig. 7.5. Right hippocampus, left putamen, and right accumbens have relatively high associations with these subnetworks.

3.2 Pathway and Network Analysis

Pathway enrichment analysis was performed to explore and analyze the functional relevance of the identified subnetworks. Several pathways, such as AD (enriched by network 6136, etc.), amyotrophic lateral sclerosis (enriched by 6682, etc.), and Huntington's disease (enriched by 1431, 2778, 4053, 6682, etc.), were identified to be related to neurodegenerative diseases. In addition, pathways of long-term

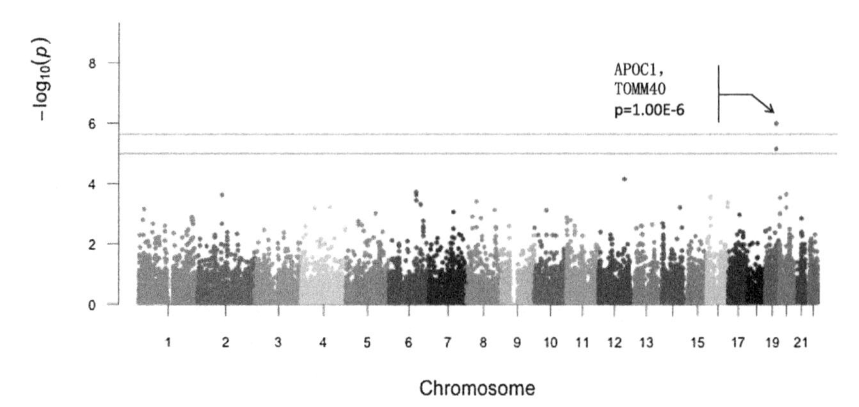

Figure 7.3 Manhattan plot of gene-based genome-wide association study (GWAS) of right hippocampus measurement. The scale of y-axis is the *P*-values (−log10(observed *P*-value)) from gene-based GWAS of right hippocampus measurement. The horizontal lines display the cutoffs for two significant levels: *blue line* (gray in print versions) for $P < 10e\text{-}6$ and *red line* (dark gray in print versions) for $P < 2.1e\text{-}6$.

depression (enriched by 1431, 2778, 6599, 6682, etc.) and long-term potentiation (enriched by 2778, etc.) were associated with nervous system. Also a lot of cancer-related pathways were identified. Many studies have focused on investigating the relationship between cancer and neurodegeneration, with abnormal cell growth, and cell loss in common [27].

For each trait prioritized subnetwork, it has some significant genes. We collect all significant genes across 14 subcortical measures, and then label each gene by the number of its significantly associated QTs. For instance, Fig. 7.6 shows an example network (i.e., #2778), where the red octangle nodes (i.e., *EGFR* and *GR1K1*) are significantly associated with 10 QTs. Both genes *EGFR* and *GR1K1* are previous demonstrated to association with

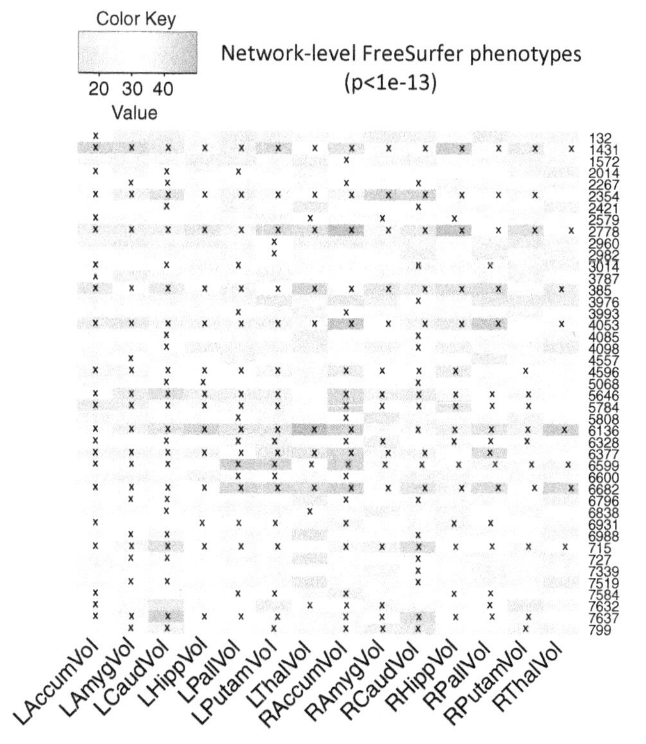

Figure 7.4 Heat map of subnetworks with subcortical quantitative traits (QTs) at the significance level of $P < 1e\text{-}13$ (uncorrected). Network-based genome-wide association study results at a statistical threshold of $P < 1e\text{-}13$ using subcortical QTs derived from FreeSurfer are shown. $-\log_{10}(P\text{-values})$ from each association are mapped and displayed in the heat map. Heat map blocks labeled with "x" reach the significance level of $P < 1e\text{-}13$. Subnetworks and QTs with at least one significant finding are included in the heat map, and thus each row or column contains at least one "x." Subnetworks *circled with red rectangle* (black in print versions) means it has significant effects across all QTs.

AD [28,29]. Also Fig. 7.7 shows another example network (i.e., #6682), where the red octangle (dark gray in print versions) node (i.e., *SHANK2*) is significantly associated with 14 QTs.

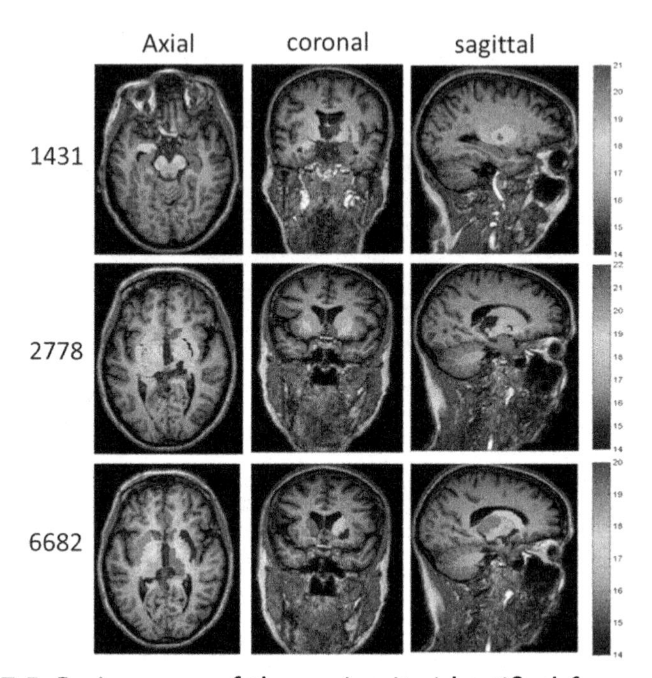

Figure 7.5 Brain maps of three circuits identified from network interface miner for multigenic interactions networks. $-\log_{10}$ (*P*-values) from each subnetwork across 14 cortical quantitative traits are mapped and displayed on the brain.

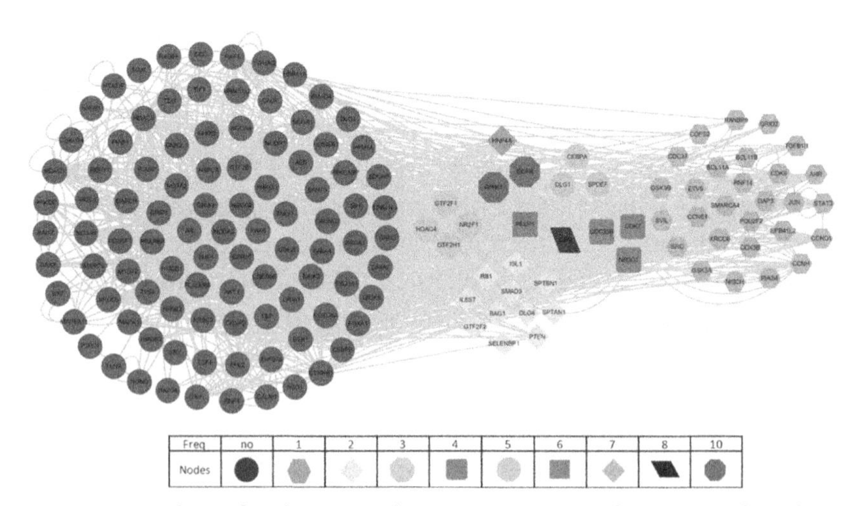

Figure 7.6 Plot of subnetwork #2778 (151 nodes in total), where each gene is labeled by the number of its significantly associated quantitative traits (QTs). For example, the *red octangle* (dark gray in print versions) nodes (i.e., *EGFR* and *GR1K1*) are significantly associated with 10 QTs.

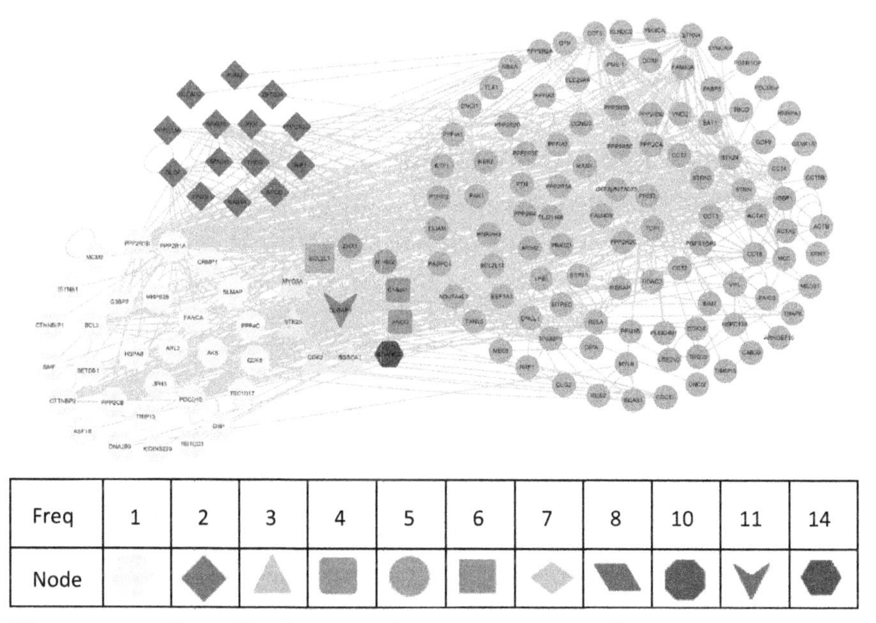

Freq	1	2	3	4	5	6	7	8	10	11	14
Node											

Figure 7.7 Plot of subnetwork #6682 (157 nodes in total), where each gene is labeled by the number of its significantly associated quantitative traits (QTs). For example, the *red octangle* (dark gray in print versions) nodes (i.e., *SHANK2*) is significantly associated with 14 QTs.

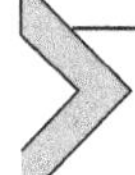

4. CONCLUSIONS

We have proposed a network-based framework, guided by PPI data, for mining high-level imaging genetic associations. We applied this framework to the ADNI data, examining multilevel (i.e., SNP, gene, and subnetwork) genetic associations with 14 subcortical measures. SNP- and gene-level analyses identified well-known AD genetic markers including *APOE* and *APOC1*. High-level subnetwork analysis can take into account information on biological relationships to interpret GWAS data and to discover trait associated subnetworks for the guidance of further study. This initial and proof-of-concept study has demonstrated that this multilevel strategy can not only replicate well-known AD genes but also identify important new genes and relevant subnetworks that are highly associated with multiple important subcortical imaging measures. Future directions include an in-depth investigation of the identified high-level patterns and replication studies in independent cohorts.

REFERENCES

[1] G.A. Blokland, et al., Genetic and environmental influences on neuroimaging phenotypes: a meta-analytical perspective on twin imaging studies, Twin Research and Human Genetics 15 (3) (2012) 351–371.

[2] L. Mosconi, et al., Early detection of Alzheimer's disease using neuroimaging, Experimental Gerontology 42 (1–2) (2007) 129–138.

[3] L. Shen, et al., Whole genome association study of brain-wide imaging phenotypes for identifying quantitative trait loci in MCI and AD: a study of the ADNI cohort, NeuroImage 53 (3) (2010) 1051–1063.

[4] J.L. Stein, et al., Genome-wide analysis reveals novel genes influencing temporal lobe structure with relevance to neurodegeneration in Alzheimer's disease, Neuroimage 51 (2) (2010) 542–554.

[5] L. Shen, et al., Genetic analysis of quantitative phenotypes in AD and MCI: imaging, cognition and biomarkers, Brain Imaging and Behavior 8 (2) (2014) 183–207.

[6] E. Rodriguez-Rodriguez, et al., Interaction between HMGCR and ABCA1 cholesterol-related genes modulates Alzheimer's disease risk, Brain Research 1280 (2009) 166–171.

[7] I. Mateo, et al., Epistasis between tau phosphorylation regulating genes (CDK5R1 and GSK-3beta) and Alzheimer's disease risk, Acta Neurologica Scandinavica 120 (2) (2009) 130–133.

[8] J. Li, Q. Zhang, F. Chen, J. Yan, S. Kim, L. Wang, W. Feng, A.J. Saykin, H. Liang, L. Shen, Genetic interactions explain variance in cingulate amyloid burden: an AV-45 PET genome-wide association and interaction study in the ADNI cohort, BioMed Research International 2015 (2015) (2015) 11.

[9] M.E.I. Koran, T.J. Hohman, T.A. Thornton-Wells, Genetic interactions found between calcium channel genes modulate amyloid load measured by positron emission tomography, Human Genetics 133 (1) (2013) 85–93.
[10] B.N. Vardarajan, et al., Rare coding mutations identified by sequencing of Alzheimer's disease GWAS loci, Annals of Neurology 78 (3) (2015) 487–498.
[11] C.M. Lill, et al., The role of TREM2 R47H as a risk factor for Alzheimer's disease, frontotemporal lobar degeneration, amyotrophic lateral sclerosis, and Parkinson's disease, Alzheimer's and Dementia 11 (12) (2015) 1407–1416.
[12] C.W. Medway, et al., ApoE variant p.V236E is associated with markedly reduced risk of Alzheimer's disease, Molecular Neurodegeneration 9 (2014) 11.
[13] C. Cruchaga, et al., Rare coding variants in the phospholipase D3 gene confer risk for Alzheimer's disease, Nature 505 (7484) (2014) 550–554.
[14] H. Shi, et al., Genetic variants influencing human aging from late-onset Alzheimer's disease (LOAD) genome-wide association studies (GWAS), Neurobiology of Aging 33 (8) (2012) 1849.e5–1849.e18.
[15] Y. Liu, et al., Gene, pathway and network frameworks to identify epistatic interactions of single nucleotide polymorphisms derived from GWAS data, BMC Systems Biology 6 (Suppl. 3) (2012) S15.
[16] S. Swaminathan, et al., Amyloid pathway-based candidate gene analysis of [(11)C]PiB-PET in the Alzheimer's Disease Neuroimaging Initiative (ADNI) cohort, Brain Imaging and Behavior 6 (1) (2012) 1–15.
[17] B. Ding, et al., Gene expression profiles of entorhinal cortex in Alzheimer's disease, American Journal of Alzheimer's Disease and Other Dementias 29 (6) (2014) 526–532.
[18] Y. Sun, et al., An integrated bioinformatics approach for identifying genetic markers that predict cerebrospinal fluid biomarker p-tau181/Abeta1-42 ratio in ApoE4-negative mild cognitive impairment patients, Journal of Alzheimer's Disease 45 (4) (2015) 1061–1076.
[19] B. Fischl, A.M. Dale, Measuring the thickness of the human cerebral cortex from magnetic resonance images, Proceedings of the National Academy of Sciences of the United States of America 97 (20) (2000) 11050–11055.

[20] S. Purcell, et al., PLINK: a tool set for whole-genome association and population-based linkage analyses, American Journal of Human Genetics 81 (3) (2007) 559–575.
[21] J.Z. Liu, et al., A versatile gene-based test for genome-wide association studies, American Journal of Human Genetics 87 (1) (2010) 139–145.
[22] N. Akula, et al., A network-based approach to prioritize results from genome-wide association studies, PLoS One 6 (9) (2011) e24220.
[23] H.H. Fu, D.K.J. Lin, H.T. Tsai, Damping factor in Google page ranking, Applied Stochastic Models in Business and Industry 22 (5–6) (2006) 431–444.
[24] D.P. Hibar, et al., Common genetic variants influence human subcortical brain structures, Nature 520 (7546) (2015) 224–229.
[25] E.Y. Chen, et al., Enrichr: interactive and collaborative HTML5 gene list enrichment analysis tool, BMC Bioinformatics 14 (2013) 128.
[26] J. Yan, et al., Hippocampal transcriptome-guided genetic analysis of correlated episodic memory phenotypes in Alzheimer's disease, Frontiers in Genetics 6 (2015) 117.
[27] K.N. Nudelman, et al., Association of cancer history with Alzheimer's disease onset and structural brain changes, Frontiers in Physiology 5 (2014) 423.
[28] Y. Hashimoto, et al., The cytoplasmic domain of Alzheimer's amyloid-beta protein precursor causes sustained apoptosis signal-regulating kinase 1/c-Jun NH_2-terminal kinase-mediated neurotoxic signal via dimerization, The Journal of Pharmacology and Experimental Therapeutics 306 (3) (2003) 889–902.
[29] C.P. Jacob, et al., Alterations in expression of glutamatergic transporters and receptors in sporadic Alzheimer's disease, Journal of Alzheimer's Disease 11 (1) (2007) 97–116.

Bayesian Feature Selection for Ultrahigh Dimensional Imaging Genetics Data

Yize Zhao[1], Fei Zou[4], Zhaohua Lu[3], Rebecca C. Knickmeyer[2], Hongtu Zhu[5]
[1]Weill Cornell Medicine, New York, NY, United States
[2]University of North Carolina at Chapel Hill, Chapel Hill, NC, United States
[3]Pennsylvania State University, State College, PA, United States
[4]University of Florida, Gainesville, FL, United States
[5]University of Texas MD Anderson Cancer Center, Houston TX, United States

Contents

Abstract

This paper is motivated by imaging genetics studies with the goal to perform feature selection among multivariate phenotypes and ultrahigh dimensional genotypes. Specifically, we propose a novel multilevel sequential selection procedure under a Bayesian multivariate response regression model to select informative features among multivariate responses and ultrahigh dimensional predictors. We treat the identification of nonzero elements in the sparse coefficient matrix into a hierarchical feature

Imaging Genetics
ISBN: 978-0-12-813968-4
http://dx.doi.org/10.1016/B978-0-12-813968-4.00008-0

selection problem by first selecting potential nonzero rows among the matrix (genotype selection) and then localizing the nonzero elements within the marked rows (phenotype selection). The genotypewise selection is accomplished by constructing multilevel auxiliary selection models under different scales with the actual scale auxiliary model treated as another level for the ultimate phenotypewise selection. We apply the method to the Alzheimer's Disease Neuroimaging Initiative with biologically meaningful results obtained.

Keywords: Bayesian feature selection; Imaging genetics; Markov chain Monte Carlo; Sequential sampling; Ultrahigh dimension

1. INTRODUCTION

In this paper, we need to build association between genetic information on more than 400,000 single-nucleotide polymorphisms (SNPs) and brain volume of 93 regions of interest (ROIs). A direct application of a multivariate response regression model (MRRM) is problematic by estimating the gigantic coefficient matrix, which motivates us to impose sparsity. Some canonical frequentist approaches including LASSO [1], elastic net [2], etc., were proposed under a univariate response regression model, and their extension to an MRRM has recently been considered. Bayesian variable selection methods are typically performed by imposing a spike and slap type of priors on the model parameters and identify the optimal model via Gibbs sampling. Alternatively, shrinkage priors can also be adopted for the model parameters with potentially more efficient computation but a lack of interpretation. In the presence of high or ultrahigh dimensional data, a few attempts have been made to overcome infeasible computation of exiting Bayesian variable selection methods [3,4]. However, these methods are not directly applicable to multivariate responses, without considering the computation could still be an issue with tens of thousands of unknown parameters.

In this paper, we develop a novel Bayesian variable selection method under an MRRM to extract informative features among both ultrahigh dimensional predictors and multivariate responses. Specifically, we transfer the two-sided selection problem on both predictors and responses into a hierarchical selection framework by first identifying informative predictors and then localizing the corresponding highly associated responses. In the posterior simulation, we construct a multilevel selection framework to transmitted the selection information from lowest level to highest level through level-depended auxiliary models, which dramatically improves the computational efficiency.

The remainder of the paper is organized as follows. In Section 2, we describe the basic model specification. In Section 3, we provide the details of our method. We conduct whole-genome whole-brain analysis in Section 4 based on Alzheimer's Disease Neuroimaging Initiative (ADNI) data and conduct with some discussion remarks in Section 5.

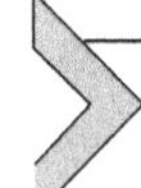

2. MODEL SPECIFICATION

We consider an imaging genetics study with n subjects in the data. For each subject, let $Y_i = (y_{i1}, \ldots, y_{iq})^{\mathrm{T}}$ denote the imaging measures, and $X_i = (x_{i1}, \ldots, x_{ip})$ denote the genetic markers. We let $\mathbf{Y} = (Y_1, \ldots, Y_n)$ and $\mathbf{X} = (X_1, \ldots, X_n)$, and we consider an MRRM for variable selection

$$\mathbf{Y} = \mathbf{X}(\mathbf{S} \circ \mathbf{B}) + \mathbf{E}, \tag{8.1}$$

where $\mathbf{B} = \left(\mathbf{b}_1^{\mathrm{T}}, \ldots, \mathbf{b}_p^{\mathrm{T}}\right)^{\mathrm{T}}$ is the coefficient matrix with $\mathbf{b}_j = (b_{j1}, \ldots, b_{jq})$ corresponding to genotype j, and $\mathbf{E} = (e_{ik})$

is the residual error matrix. Denoting vector $\mathbf{e}_i = (e_{i1},\ldots, e_{iq})$, we assume $\mathbf{e}_i \sim \mathrm{N}(0,\Sigma)$ with Σ the covariance matrix. Under the entrywise product "$\circ$", we introduce logical matrix $\mathbf{S} = (s_{jk})$ with $s_{jk} \in \{0,1\}$ indicating the joint selection status for both genotypes j and phenotypes k. As a direct extension from that in univariate response case, posterior inference could be conducted via Gibbs sampling in model (1). However, performing such a procedure is indeed unrealistic given the total number of unknown parameters involved is at least $2pq + q(q + 1)/2$, which can be huge with even moderate p and q (more than 9×10^7 in our application). Besides the prohibitive computation, the feature selection performance will also be deteriorate given a considerably smaller sample size.

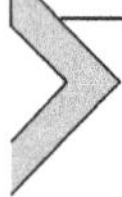

3. MULTILEVEL BAYESIAN FEATURE SELECTION FRAMEWORK

In this section, we develop a Bayesian feature selection method under multilevel framework to address challenges to fit model (1). We first construct G levels of partitions for all the p genotypes with partition g dividing the whole genome into p^g subsets. The number of predictors in subsets s ($s = 1, \ldots, p^g$) is J_s^g. Thus, we have $p^1 < p^2 < \ldots, <p^G = p$ and $\sum_{s=1}^{p^g} J_s^g = p$, for $g = 1, \ldots, G$. The partitions between adjacent levels are nested with each other, and the subsets within each partition are mutually exclusive. Specifically, we introduce two logical matrices $\mathbf{U}^g$ and $\mathbf{V}^g$ at level g, where $\mathbf{U}^g = \left(u_{sj}^g\right)$ with $u_{sj}^g = 1$ if predictor j belongs to subset s; and $\mathbf{V}^g = \left(v_{st}^g\right)$ with $v_{st}^g = 1$ if subset t at level g belongs to subset s at level $g - 1$, and we define $\mathbf{V}^1 = \mathbf{I}$. In the real application, among genetic markers, one can

construct the subsets of SNPs based on their geometric location along the genome, or in light of SNP-sets information.

We introduce selection indicator under level g as $D^g = \left(d_1^g, \ldots, d_{p^g}^g\right)$. To build selection model for the p^g subsets at level g, we divide $\mathbf{X}$ into $\left(\mathbf{X}_1^g, \ldots, \mathbf{X}_{p^g}^g\right)$ with submatrix $\mathbf{X}_s^g$ corresponding to subset s. Afterward, we adopt a reduced rank singular-value decomposition $\mathbf{X}_s^g = \mathbf{Z}_s^g \mathbf{A}_s^g$, for $s = 1, \ldots, p^g$. Based on $\mathbf{Z}^g = \left(\mathbf{Z}_1^g, \ldots, \mathbf{Z}_{p^g}^g\right)$, after adjusting mapping matrix $\mathbf{U}^g$, our alternative auxiliary model becomes

$$\mathbf{Y} = \mathbf{Z}^g \left\{ \left(\mathbf{U}^{g\mathrm{T}} \cdot D^g \cdot \mathbf{1}_q^{\mathrm{T}} \right) \circ \mathbf{W}^g \right\} + \mathbf{E}^g, \tag{8.2}$$

where matrix $\mathbf{W}^g = \left(\mathbf{w}_1^g, \ldots, \mathbf{w}_{L^g}^g\right)$ is the coefficients matrix based on factors. In practice, it is desirable to impose certain truncation to each L_s^g, in particular at lower levels. By doing so, model (2) realizes a dimension reduction of the cardinality of predictors from p to L^g at level g. At the highest level, we work on the selection on actual scale predictors.

We introduce priors for the parameters in the level-specific internal selection model. Specifically, for model (2), we assign

$$\begin{aligned} &\mathbf{w}_j^{(g)} \sim \mathrm{N}_q\left(0, \mathbf{I}_q \sigma_j^{(g)}\right), \quad \text{with } \sigma_j^{(g)} \sim IG(\alpha_1, \beta_1); \\ &\text{for } j = 1, \ldots, L^g; \; g = 1, \ldots, G; \end{aligned} \tag{8.3}$$

$$d_k^{(g)} \sim \mathrm{Bern}(\pi_g); \quad \text{for } k = 1, \ldots, p^g; \; g = 1, \ldots, G. \tag{8.4}$$

For the covariance matrix $\Sigma^{(g)}$, we redefine it as a diagonal matrix $\Sigma^{(g)} = \mathrm{Diag}\left(\tau_1^{(g)}, \ldots, \tau_q^{(g)}\right)$ and $\tau_k^{(g)} \sim IG(\alpha_2, \beta_2)$.

3.1 Multilevel Sequential Sampling Procedure

To construct a valid posterior sampling procedure, we first introduce an auxiliary subset selection indicator $\widetilde{D}^{g-1} = \left(\widetilde{d}_1^{g-1}, \ldots, \widetilde{d}_{p^{g-1}}^{g-1}\right)$ defined as $\widetilde{d}_s^{g-1} = \max\left\{d_t^g : v_{st}{}^g = 1\right\}$. Indicator $\widetilde{D}^{g-1}$ defined at level g has the same dimension and structure as the selection indicator D^{g-1} at level $g-1$.

The posterior distribution under the auxiliary model at level g is

$$\pi\left(\mathbf{W}^g, D^g, \widetilde{D}^{g-1}, \boldsymbol{\sigma}^g, \boldsymbol{\tau}^g \middle| \cdot\right) = \pi(\mathbf{W}^g, D^g, \boldsymbol{\sigma}^g, \boldsymbol{\tau}^g | \cdot)\pi\left(\widetilde{D}^{g-1} \middle| D^g\right),$$

The first term in the right-hand side of Eq. (8.5) is the main probability part, which indicates the updating scheme for parameters $\{\mathbf{W}^g, \boldsymbol{\sigma}^g, \boldsymbol{\tau}^g\}$ consistent along different levels. Parameters $\{\mathbf{W}^g, \boldsymbol{\sigma}^g, \boldsymbol{\tau}^g\}$ can be updated via Gibbs sampler. And we adopt a Metropolis–Hastings (M–H) adjusted step followed by a Gibbs sampler moving step to jointly update selection indicator and auxiliary indicator $\left\{D^g, \widetilde{D}^{g-1}\right\}$. Specifically, In the M–H step, we design the following proposal distribution

$$f\left\{\left(D_c^g, \widetilde{D}_c^{g-1}\right) \rightarrow \left(D_*^g, \widetilde{D}_*^{g-1}\right) \middle| \cdot\right\}$$
$$= H\left(D_*^g \middle| \widetilde{D}_*^{g-1}, D_c^g, \widetilde{D}_c^{g-1}\right) P_g\left(\widetilde{D}_*^{g-1} \middle| \mathbf{Y}, \mathbf{X}\right), \qquad (8.5)$$

For Eq. (8.5), function $P_g(\cdot)$ specifying the sampling scheme for auxiliary indicator $\widetilde{D}^{g-1}$ is designed to only depend on the data. Without the interference of parameters in the current level, $P_g(\cdot)$ can be approximated by the posterior samples of selection indicator D^{g-1} in the auxiliary model at level $g-1$. Given the sampled value of $\widetilde{D}^{g-1}$ and the current state of Markov chain, function $H(\cdot)$ specifics the sampling scheme for the current level selection indicator D^g, which allows the transformation of selection information from previous level to the current level.

3.2 Hierarchical Variable Selection

We further treat the model at level G as another auxiliary model that can further induce the sampling procedure of hierarchical variable selection of the target model (1). Following the sampling procedure stated in the previous sections, we introduce priors for parameters in the target model

$$\mathbf{b}_j \sim N_q(0, \mathbf{I}_q\sigma_j), \text{ with } \sigma_j \sim IG(\nu_1, \eta_1); \quad \text{for } j = 1, \ldots, p;$$
$$s_{jk} \sim \text{Bern}(\rho); \quad \text{for } j = 1, \ldots, p; \ k = 1, \ldots, q; \tag{8.6}$$

For the covariance matrix, we decompose random error $\mathbf{e}_i$ based on an infinite Bayesian factor model $\mathbf{e}_i = \Lambda\mathbf{l}_i + \mathbf{t}_i$ with factor loading matrix Λ, latent factor $\mathbf{l}_i \sim \mathrm{N}_\infty(0, I_\infty)$ and $\mathbf{t}_i \sim \mathrm{N}(0, \Sigma_t)$ with $\underset{t}{\Sigma} = \mathrm{Diag}(\delta_1, \ldots, \delta_q)$. We further place a multiplicative gamma process shrinkage prior proposed by [5] on factor loading matrix Λ to shrink its elements to zero with the increase of column index.

We introduce auxiliary indicator $\widetilde{D} = \left(\widetilde{d}_1, \ldots, \widetilde{d}_q\right)$ with $\widetilde{d}_k = \max\{s_{jk}, \ j = 1, \ldots, p\}$ for $k = 1, \ldots, q$ for the final model to borrow genotype selection information from the genotype selection model (auxiliary model at level G). Accordingly, the posterior distribution becomes

$$\pi\left(\mathbf{B}, \mathbf{S}, \widetilde{D}, \boldsymbol{\sigma}, \boldsymbol{\tau} \middle| \mathbf{Y}, \mathbf{X}\right) = \pi(\mathbf{B}, \mathbf{S}, \boldsymbol{\sigma}, \boldsymbol{\tau} | \mathbf{Y}, \mathbf{X}) \pi\left(\widetilde{D} \middle| \mathbf{S}\right), \tag{8.7}$$

where sampling procedure closely follows that in Section 3.1. Based on the marginal posterior inclusion probability of each element in **S**, we can summarize the informative phenotype–genotype pairs.

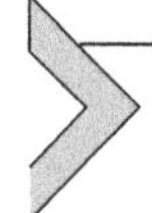

4. ALZHEIMER'S DISEASE NEUROIMAGING INITIATIVE

We conduct analysis based on the neuroimaging, genetics, and clinical data under the ADNI study, and our goal is to build association between SNPs and imaging markers and select information association pairs. For imaging traits, after preprocessed the raw MRI data by standard steps [6], 93 ROIs were labeled, and the volumes of all the ROIs are calculated for each subject. A total of 745 subjects with 421,823 SNPs (16,084 SNP-sets) on each subject are left in our analysis, and we also include age, intracranial volume (ICV), gender, education, and handedness as covariates.

We apply the proposed multilevel-based method to the data. In light of the SNP-set information, we deliver the posterior inference for the genotype selection into five levels with the selection unit of the fourth level as SNP-set and reach the final selection on both genotypes and phenotypes subsequentially. For the final model and auxiliary models, the Markov chain Monte Carlo algorithm is implemented with 10,000 iterations and 5000 burn-in with the convergence checked by Gelman–Rubin diagnosis.

We summarize the top 12 selected SNPs along with the location of ROIs that are associated with each information genotypes in Fig. 8.1. The size of the nodes is represented by the posterior inclusion probability of the genotype–phenotype selection indicator, and we only include the nodes with a corresponding posterior probability larger than 0.1. From the informative genotypes, rs11707680 and rs11711797 are located in gene ST6GAL1, which is a famous T2 diabetes risk factor, and recent research suggests a higher incidence of cognitive decline and an increased risk of developing dementia for diabetes. SNPs rs2074634 and rs346535 located in gene SMG9 regulate remodeling of the mRNA surveillance complex during nonsense-mediated mRNA decay. They are highly associated with the left side of lateral ventricle in the brain. Genetic marker rs2075650 in gene ApoE4 is the one with high associations with multiple brain regions including both sides of

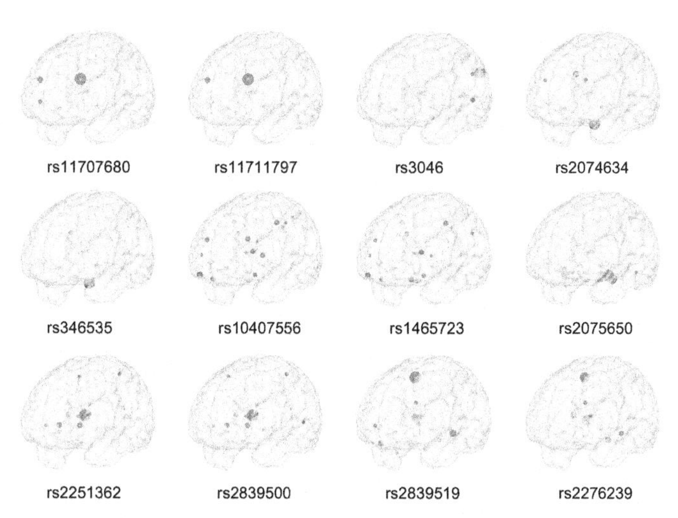

Figure 8.1 The top 12 selected genotypes and their associated regions of interest. The size of the nodes represents the marginal posterior probability of the selection indicator.

hippocampal formation, amygdala, and parahippocampal gyrus. As it is well known, ApoE4 is the largest known genetic risk factor for Alzheimer's Disease, and our results are consist with this knowledge with further localizing the specific ROIs associated with this biomarker.

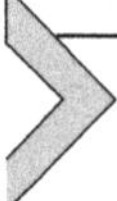

5. DISCUSSION

In this article, we present a novel multilevel sequential selection procedure under a Bayesian MRRM to select informative features among multivariate responses and ultrahigh dimensional predictors. Although our method is implemented by specific partitions on predictors and auxiliary selection models, it can be readily modified with a different partition scheme or auxiliary models. For instance, we could construct our partition based on the pairwise correlation between predictors and responses. For the internal selection procedure, we currently use an *i.i.d* Bernoulli prior for the selection indicator. To incorporate more biological information in the selection procedure, we could estimate the underlying biological network among genotypes and functional connectivity among phenotype, and further utilize Markov random field/Ising prior to model such information. Our method is currently applied to the study with a dimension of response up to 100. To further extend it, to handle higher or even ultrahigh dimensional response, we could construct parallel multilevel selection procedure on the response level selection as well. One potential issue of such approach is that the variable selection performance of the auxiliary models might be fragile because of a coarse-scale selection on both genotypes and phenotypes. As a result, the choice of auxiliary models will need to be further investigated and we treat this as a future work.

REFERENCES

[1] R. Tibshirani, Regression shrinkage and selection via the lasso, Journal of the Royal Statistical Society: Series B (Methodological) (1996) 267–288.
[2] H. Zou, T. Hastie, Regularization and variable selection via the elastic net, Journal of the Royal Statistical Society: Series B (Statistical Methodology) 67 (2) (2005) 301–320.
[3] V.E. Johnson, D. Rossell, Bayesian model selection in high-dimensional settings, Journal of the American Statistical Association 107 (498) (2012) 649–660.
[4] V.E. Johnson, On numerical aspects of Bayesian model selection in high and ultrahigh-dimensional settings, Bayesian Analysis 7 (4) (2013) 1–18.
[5] A. Bhattacharya, D.B. Dunson, et al., Sparse Bayesian infinite factor models, Biometrika 98 (2) (2011) 291.
[6] D. Shen, C. Davatzikos, Measuring temporal morphological changes robustly in brain MR images via 4-dimensional template warping, NeuroImage 21 (4) (2004) 1508–1517.

CHAPTER NINE

Continuous Inflation Analysis: A Threshold-Free Method to Estimate Genetic Overlap and Boost Power in Imaging Genetics

Derrek P. Hibar[1], Neda Jahanshad[1], Sarah E. Medland[2], Paul M. Thompson[1]
[1]Keck School of Medicine of USC, Marina del Rey, CA, United States
[2]QIMR Berghofer Medical Research Institute, Brisbane, QLD, Australia

Contents

Imaging Genetics
ISBN: 978-0-12-813968-4
http://dx.doi.org/10.1016/B978-0-12-813968-4.00009-2

Abstract

Methods to quantify genetic overlap may elucidate relationships between disparate traits and provide Bayesian priors to guide the search for genetic influences on brain measures. Here we describe a threshold-free method called continuous inflation analysis, which we used to compare genome-wide association statistics (GWAS) for the volumes of eight brain regions, computed from brain magnetic resonance imaging. Our goal was to understand the extent of pleiotropy (overlap in genetic influences) and concordance for the volumes of brain regions with different biological functions. We found significant pleiotropy among seven of the subcortical brain volumes. We found positive concordance across the seven subcortical structures and negative concordance between genetic influences on each subcortical structure and intracranial volume. Using a conditional false discovery rate approach, we showed that a given brain volume GWAS could act as a Bayesian prior and improve the power to detect novel associations in a related brain volume. When conditioning the putamen volume GWAS on the caudate volume GWAS, we identified 17 novel loci associated with putamen volume.

Keywords: Concordance; Genetic correlation; Genetic overlap; Neuroimaging; Genetics; Pleiotropy

1. INTRODUCTION

Recent imaging genetics work in the Enhancing NeuroImaging Genetics through Meta-Analysis (ENIGMA) Consortium has focused on discovering common genetic variants associated with the volumes of seven subcortical brain structures (nucleus accumbens, amygdala, caudate, hippocampus, globus pallidus, putamen, thalamus) and one measure of global head size (intracranial volume; ICV) [1]. Hibar et al. examined individual single-nucleotide polymorphism (SNP) associations with each of the eight brain volumes—each considered as a single trait—but did not examine the overlapping genetic influence of the full set of common variants across structures. By examining the pleiotropy (common genetic influences) and concordance[1] across subcortical structures we should be able to (1) define Bayesian approaches to guide the search for genetic influences on the brain and (2) better understand the underlying genetic pathways that may partially explain volumetric variations across different brain regions.

Twin and family studies can estimate the *genetic correlation* between subcortical brain volume traits [2], i.e., the fraction of the observed correlation that is due to genetic factors. However, there may be little or no publicly available twin or family data for a given pair of traits. If genome-wide SNP data are available for a cohort, we can use genome-wide association statistics (GWAS) summary statistics (i.e., regression coefficients relating each SNP to the traits of interest) to estimate the common genetic overlap. This is perhaps surprising because almost all SNPs have no detectable effects and even significantly associated SNPs generally have weak effects. So, without vast samples of data, it can be challenging to pick up genetic overlap from SNP association data. A recent method called linkage disequilibrium score (LDSC) regression [3] uses GWAS summary statistics from two traits to estimate a genetic correlation driven by common genetic determinants. One limitation of this method (and genetic correlations calculated

[1] Concordance is an extension to the concept of pleiotropy that includes the *direction* of a pleiotropic SNP effect (i.e., a positive or negative correlation).

from twin and family studies as well) is that it is not possible to identify which *specific* variants overlap and contribute to the correlation. A related method, SNP effect concordance analysis [4], looks at pleiotropy and concordance and predefined, arbitrary thresholds. Even so, it is of great interest to try to narrow the search for genetic variants associated with brain measures, to avoid heavy multiple comparisons corrections and the vast sample sizes they currently imply (often requiring tens of thousands of subjects, e.g., in the ENIGMA studies).

Here we describe a novel method to quantify the global enrichment (pleiotropy) and concordance between GWAS summary statistics from two traits. We apply this method to examine the genetic overlap between brain structures examined in the ENIGMA Consortium. Our hypothesis is that brain regions will show genetic overlap with structures similar to their functional groupings: limbic system (hippocampus, amygdala, thalamus) and basal ganglia (putamen, caudate, nucleus accumbens, and globus pallidus). Furthermore, we examine whether a conditional false discovery rate (FDR) framework can be used to boost power to detect novel associations.

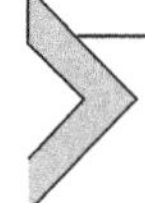

2. METHODS

2.1 Estimating the Genetic Overlap Between Two Traits

We developed a data-driven, threshold-free method, called continuous inflation analysis (CIA), to assess global enrichment (pleiotropy) and concordance based on GWAS summary statistics from any two pairwise traits. Here we were interested in assessing the genetic overlap across the volumes of eight different brain regions: the nucleus accumbens, amygdala, caudate, hippocampus, globus pallidus, putamen, thalamus, and ICV. These data were obtained from the discover cohort of the ENIGMA Consortium [1] and are composed of 13,171 subjects of European descent. The GWAS was performed while controlling for age, age^2, sex, four multidimensional scaling (MDS) components, ICV (for non-ICV phenotypes), and diagnosis (when applicable). In cases where family data were included an appropriate mixed-effects model (controlling for relatedness) was performed before inclusion in the final metaanalyzed sample. We performed all pairwise combinations of overlap tests between the eight traits.

Before comparing two traits, we designated one data set the *reference* data set and the other the *test* data set. This designation is important because the CIA procedure is not symmetric. To begin, we performed a clumping procedure to select independent index SNPs for each LD block in the genome. The index SNPs were chosen based on significance levels in the *reference* data set (PLINK options: –clump-p1 1 –clump-p2 1 –clump-kb 500 –clump-r2 0.2) [5]. This corresponds to a *P*-value inclusion threshold and index SNP *P*-value both equal to 1, which allows for the selection of the most significant SNP within a given LD block. The last two PLINK parameters designate that we searched for pairwise relationships between SNPs (to estimate LD blocks) within a 500 kilobase window, and two SNPs were considered to be related if the R^2 between them >0.2. Next, we merged the *reference* and *test* data set such that only GWAS summary statistics for the index SNPs remained in the data set. We estimated the global enrichment (pleiotropy) by first sorting the merged data set by the *P*-value of each SNP in the *reference* data set (in descending order). We iterated through the sorted merged data set at a given step size ($n = 100$), with each step moving down the list by n SNPs. At each step, we calculated amount of enrichment by comparing the empirical cumulative distribution function (ecdf) of *P*-values from the *test* data set for SNPs from *n* to the end of the list with the ecdf of the full set of *P*-values from the *test* data set. For example, say you are looking at the subset of SNPs in your *test* data set where the subset is chosen such that it only includes SNPs with *P*-value $< .05$ in your *reference* data set. Taking the ecdf of the subsetted set of SNPs and *P*-values from the *test* data set you can determine if the set deviates from a null distribution. In this case the *null* distribution is the full set of SNPs and *P*-values from the *test* data set without subsetting. Leftward deflections in the subsetted ecdf were considered evidence of enrichment at a given cutoff and were estimated using a one-sided, two-sample Kolmogorov–Smirnov (KS) test. The comparison of two traits was considered to have significant evidence of pleiotropy if the *P*-value vector over all cutoffs exceeded the Benjamini–Hochberg false discovery rate [6] (BH-FDR) threshold (set to $q = (0.05/8 \text{ traits}) = 0.00625$). A description of the CIA workflow is given in Fig. 9.1.

As an extension to the pleiotropy tests, we performed a test of *concordance*, which considers the effect direction (the sign of the beta-coefficient from the regression of a given SNP against a given trait) when assessing the extent of overlap between two traits. The concordance can be negative (the effect direction in the test data set is negative when the effect direction is positive in the reference data set, and vice versa), positive (the effect direction is positive or negative in both data sets) or null when there is no evidence for concordance. The concordance test can be simply applied by filtering out SNPs from the merged data set (keeping either negative or positive concordant SNPs) and then continuing on with the CIA procedure described above. The significance is calculated in the same way (with the KS test) and overall global evidence of concordance was determined over all cutoffs (at BH-FDR $q = 0.00625$).

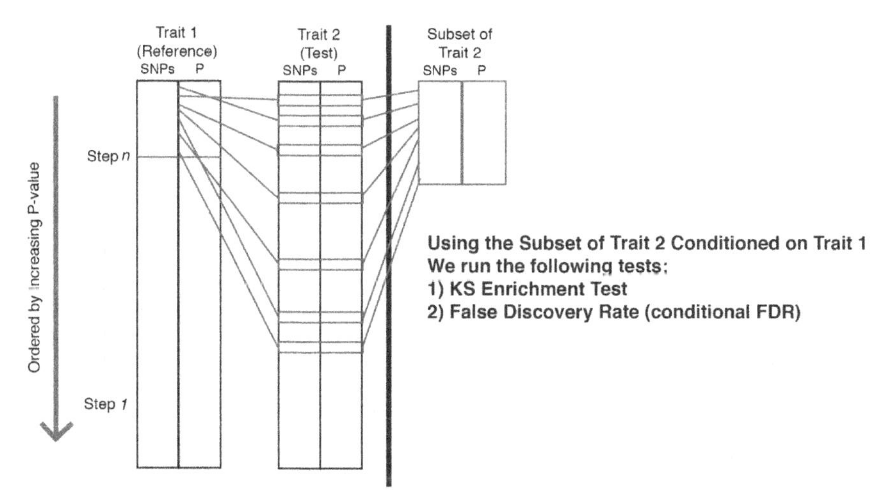

Figure 9.1 Workflow diagram of continuous inflation analysis for a given step. In this case we are conditioning the genome-wide association statistics (GWAS) results from Trait 2 (the test data set) on the GWAS results from Trait 1 (the reference data set) to obtain a new subset of Trait 2 and testing whether or not this new subset is more enriched with significant single-nucleotide polymorphisms (SNPs) than can be expected by chance. We save the *P*-value of enrichment over all possible cut-offs (Step *n*) and run Benjamini–Hochberg false discovery rate (BH-FDR) to assess significance. Finally, within the most significant subset of Trait 2 we run BH-FDR to see if additional SNPs can be declared significant compared with running BH-FDR on the full GWAS results for Trait 2. *KS*, Kolmogorov–Smirnov.

2.2 Examining Bias in Enrichment Tests Using a Negative Control

We obtained GWAS summary statistics from a skin-based trait (presence of a whorl on the left thumb [7]) to provide a negative control for our enrichment tests. The presence of whorls in fingerprints is unlikely to be related to brain volume phenotypes (and no previous link has been made in the literature) so estimating genetic overlap between brain volume GWAS and fingerprint whorl GWAS can provide evidence of Type II error bias in the CIA enrichment test model. Fingerprint data were collected from rolled ink prints and manually examined at the Queensland Institute of Medical Research and is described elsewhere [7]. The fingerprint whorl GWAS was based on data from 3314 participants (twins and their family members) using genotypes imputed to the 1000 Genomes phase 1, version 3 reference panel [8].

2.3 Boosting Power to Detect Novel Gene Variants Using Conditional False Discovery Rate

For pairwise comparisons that show significant overlap, we can boost the power to detect individual SNPs associated with a given test trait by conditioning on the reference GWAS data set. From the CIA model for a given pairwise comparison, we can choose the step-based cutoff that results in the most significant enrichment over all possible cutoffs. Next, we can apply the BH-FDR to the SNP, *P*-values from the subsetted test data set with $q = 0.05$. For comparison, we applied the BH-FDR to the full set of SNP, *P*-values from the test data set with $q = 0.05$. SNPs that pass BH-FDR in the subsetted data set but not in the full data set are considered to be detected with increased power when conditioning on the reference data set.

3. RESULTS

3.1 Pleiotropic Gene Variants Influence Multiple Brain Regions

We found significant evidence for pleiotropy between all pairwise comparisons of seven subcortical brain volumes (see Fig. 9.2). None of the pairwise comparisons with ICV

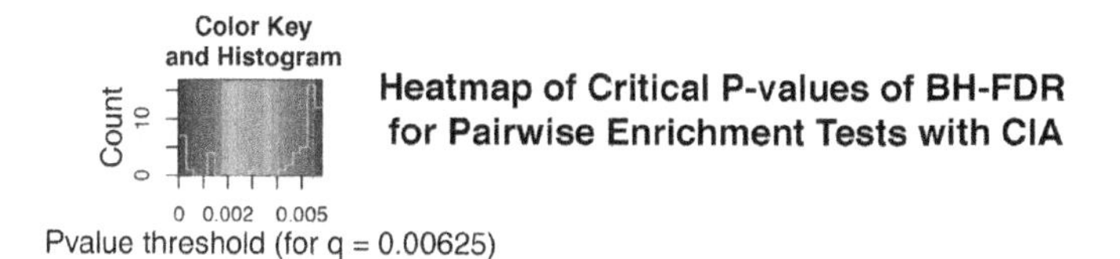

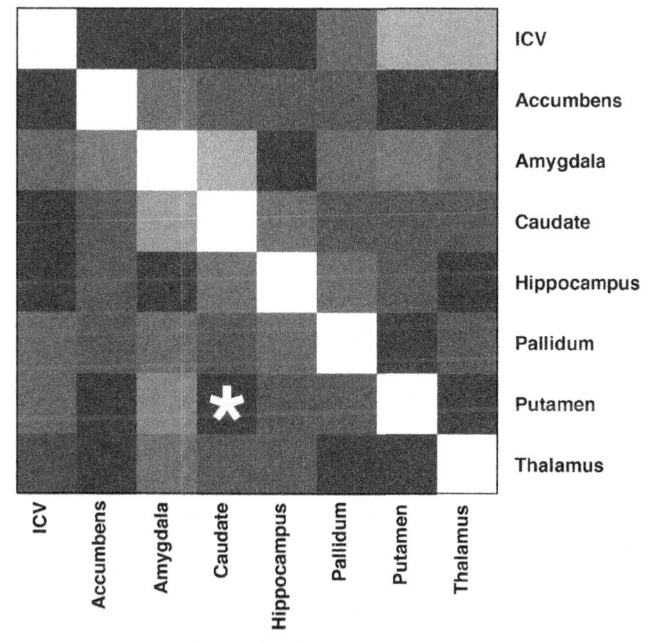

Figure 9.2 Global evidence of pleiotropy for pairwise comparisons of eight brain traits. Comparisons were made using continuous inflation analysis (CIA) and were considered significant at a Benjamini–Hochberg false discovery rate (BH-FDR) threshold ($q = 0.00625$). The seven subcortical brain structures were tightly linked in terms of pleiotropy, but no structures showed evidence of pleiotropy with intracranial volume (ICV). The most significant comparison (Putamen | Caudate) is marked with a *white star*.

showed significant overlap. The most significant comparison showing the highest evidence of pleiotropy occurred between the putamen and caudate ($q = 0.0058$). This relationship makes intuitive sense given the strong functional and histological evidence linking the two basal ganglia brain structures together.

3.2 Evidence of a Positive Concordance Between Subcortical Brain Structures

We found significant positive concordance in each of the pairwise comparisons of subcortical brain traits (see Fig. 9.3). In other words, genetic variants associated with an increase in a given brain volume also tend to be

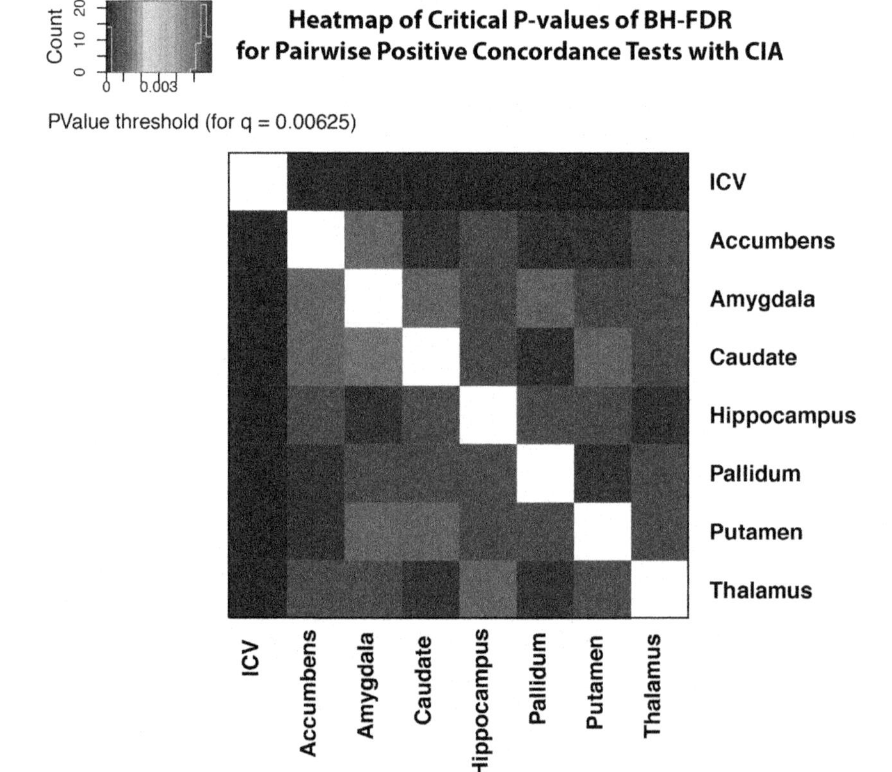

Figure 9.3 Global evidence of positive concordance for pairwise comparisons of eight brain traits. Comparisons were made using continuous inflation analysis (CIA) and were considered significant at a Benjamini–Hochberg false discovery rate (BH-FDR) threshold ($q = 0.00625$). The seven subcortical brain structures were tightly linked in terms of positive concordance, whereas none of the structures showed evidence of positive concordance with intracranial volume (ICV).

associated with an increase in the volume of another subcortical trait (and vice versa). Here there is no detectable evidence of positive concordance between the subcortical brain structures and ICV.

We found significant negative concordance in each of the pairwise comparisons between ICV and subcortical traits (see Fig. 9.4). In other words, gene variants that are

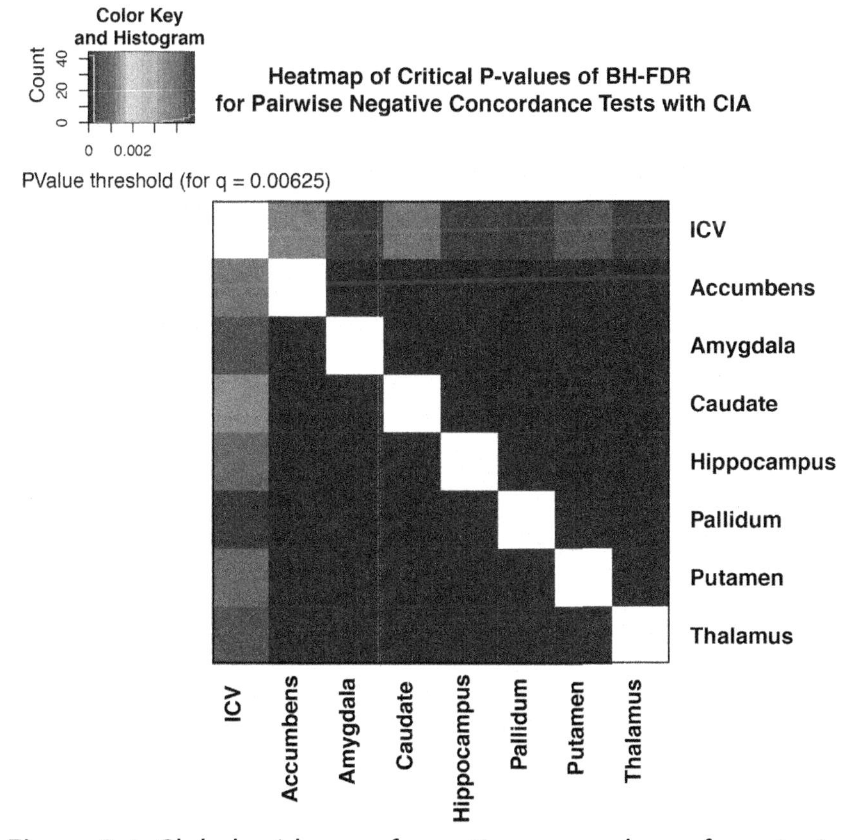

Figure 9.4 Global evidence of negative concordance for pairwise comparisons of eight brain traits. Comparisons were made using continuous inflation analysis (CIA) and were considered significant at a Benjamini–Hochberg false discovery rate (BH-FDR) threshold ($q = 0.00625$). The pairwise comparisons with intracranial volume (ICV) and the other seven structures showed significant negative concordance, whereas the pairwise comparisons among the subcortical brain volume traits were not significant.

associated with an increase in ICV also tend to be associated with a *decrease* in subcortical brain volume. However, it is worth noting that the subcortical volume GWASs were corrected for ICV as a linear predictor so the relationship here likely represents any residual nonlinear relationship. In general, there is a positive phenotypic correlation between subcortical volumes and ICV.

3.3 Finger Whorl Pattern as a Negative Control for Enrichment Tests in Brain

We found no evidence of pleiotropy between putamen volume and the dermatoglyphic negative control (presence of whorl on the left thumb) at an FDR g-value = 0.05.

3.4 Conditioning Enrichment Tests on Another Brain Prior Can Boost Power to Detect Effects in the Original Trait

Several of the pairwise comparisons of pleiotropy were significant, so, for purposes of illustration of the method, here we give the conditional FDR results for the "most significant" comparison (putamen volume GWAS conditioned on caudate volume GWAS). We identified 17 additional significant variants influencing putamen volume that were previously undetected without conditioning on the caudate volume GWAS (see Table 9.1).

4. CONCLUSIONS

We discovered evidence of significant pleiotropy between gene variants influencing different subcortical brain volumes, using CIA. This agrees with findings from twin and family heritability studies, which show that there is significant genetic correlation for volumetric measures

Table 9.1 Conditional false discovery rate (FDR) analysis of putamen genome-wide association statistics (GWAS) conditioned on caudate GWAS

SNP	Raw *P*-value in putamen GWAS	Subset FDR	FDR of full sample
rs4888010	4.92E-07	0.025	0.071
rs11150623	1.25E-06	0.027	0.076
rs17388257	1.41E-06	0.027	0.076
rs62394265	1.42E-06	0.027	0.076
rs76647989	7.70E-07	0.027	0.076
rs7873504	1.37E-06	0.027	0.076
rs10963102	2.77E-06	0.029	0.080
rs12487861	3.16E-06	0.029	0.080
rs184917581	3.10E-06	0.029	0.080
rs6135525	3.04E-06	0.029	0.080
rs62022639	2.12E-06	0.029	0.080
rs6869844	2.26E-06	0.029	0.080
rs7325851	2.47E-06	0.029	0.080
rs80258284	2.30E-06	0.029	0.080
rs842389	1.92E-06	0.029	0.080
rs10033333	4.72E-06	0.041	0.113
rs115186168	5.00E-06	0.041	0.113

Shown here are variants that pass FDR at $q = 0.05$ in the putamen volume GWAS when prioritizing single-nucleotide polymorphisms (SNPs) based on their significance in the caudate GWAS, but do not pass FDR when considering the Full set of putamen GWAS variants.

of the subcortical structures [2]. The CIA analysis builds on the twin and family heritability estimates because the overlap between traits is estimated from GWAS only and does not require a family or twin design—it can be applied to imaging genetic studies of unrelated individuals, which are more common. The most significant evidence of pleiotropy came from the putamen volume GWAS conditioned on the caudate volume GWAS. The close relationship between gene variants effecting both structures is intuitively reasonable, given the histological similarity of the caudate and putamen tissue [9].

Besides the known relationships, it appears that gene variants that explain variances in the volumes of subcortical structures that were previously thought to be independent do indeed have an effect (in fact, all subcortical structures showed a significant relationship with all other subcortical structures). The lack of enrichment between ICV and subcortical structures is not surprising, given that the subcortical volume GWAS is controlled for ICV [1]. Curiously though, we find evidence of *negative* concordance between ICV and subcortical structures. This is likely due to nonlinear differences in ICV that are not fully accounted when adjusting subcortical brain volume GWAS with ICV as a linear predictor [10]. Among subcortical structures, we found that there is a *positive* concordance (but no effect when compared with ICV). Subcortical GWAS without correction for ICV is not available at this time so we are not able to disentangle the unique contributions of a global correction for head size with ICV.

One distinct advantage of pairwise comparisons of traits with CIA is the ability to identify specific SNPs with pleiotropic effects. An extension of this idea is then to use the GWAS of a trait as part of a Bayesian prior for a related trait, to boost the power to detect effects. When looking at the comparison with the most significant evidence of pleiotropy (Putamen | Caudate) with a conditional FDR approach, we were able to identify 17 additional significant loci. The top loci (rs4888010) are an intergenic SNP on chromosome 16q22.3 [11], but further analysis of this and the other 16 loci is necessary, to better understand potential mechanisms that may influence putamen volume. All of these models are performed in the context of common genetic variants commonly known as SNPs. It is likely the case that further contributions to genetic overlap common from other forms of genetic variation such as copy number variants or insertions/deletions.

Applying CIA to other traits including those involving neuropsychiatric disease risk will help to quantify the genetic overlap between brain-related phenotypes and brain disorders and may provide a cost-effective method to screen potential endophenotypes with existing data. Furthermore, CIA combined with conditional FDR may identify new susceptibility loci for neuropsychiatric disease risk that would have previously been undetected.

ACKNOWLEDGMENTS

ENIGMA was supported in part by a Consortium grant (U54 EB020403 to PMT) from the NIH Institutes contributing to the Big Data to Knowledge (BD2K) Initiative, including the NIBIB and NCI.

REFERENCES

[1] D.P. Hibar, et al., Common genetic variants influence human subcortical brain structures, Nature 520 (7546) (2015) 224–229.
[2] M.E. Rentería, et al., Genetic architecture of subcortical brain regions: common and region-specific genetic contributions, Genes, Brain and Behavior 13 (8) (2014) 821–830.
[3] B.K. Bulik-Sullivan, et al., LD Score regression distinguishes confounding from polygenicity in genome-wide association studies, Nature Genetics 47 (3) (2015) 291–295.
[4] D.R. Nyholt, SECA: SNP effect concordance analysis using genome-wide association summary results, Bioinformatics (2014), http://dx.doi.org/10.1093/bioinformatics/btu171.
[5] S. Purcell, et al., PLINK: a tool set for whole-genome association and population based linkage analyses, The American Journal of Human Genetics 81 (3) (2007) 559–575.

[6] Y. Benjamini, Y. Hochberg, Controlling the false discovery rate: a practical and powerful approach to multiple testing, Journal of the Royal Statistical Society: Series B (Methodological) (1995) 289–300.
[7] Y.Y.W. Ho, et al., Genetic variant influence on whorls in fingerprint patterns, The Journal of Investigative Dermatology 136 (4) (2016) 859.
[8] 1000 Genomes Project Consortium, An integrated map of genetic variation from 1,092 human genomes, Nature 491 (7422) (2012) 56–65.
[9] J. Yelnik, et al., A three-dimensional, histological and deformable atlas of the human basal ganglia. I. Atlas construction based on immunohistochemical and MRI data, NeuroImage 34 (2) (2007) 618–638.
[10] C.C. Brun, et al., Sex differences in brain structure in auditory and cingulate regions, NeuroReport 20 (10) (2009) 930.
[11] http://www.broadinstitute.org/mammals/haploreg/haploreg.php.

INDEX

'*Note*: Page numbers followed by "f" indicate figures, "t" indicate tables, and "b" indicate boxes.'

D

E

F

G

H

I

K

L

M

N

O

P

Q

R

S

Printed in the United States
By Bookmasters